THERAPEUTIC TOUCH

THERAPEUTIC TOUCH
Research, Practice and Ethics

EDITED BY
Martin Rovers, Judith Malette and Manal Guirguis-Younger

University of Ottawa Press
2017

uOttawa

The University of Ottawa Press gratefully acknowledges the support extended to its publishing list by Canadian Heritage through the Canada Book Fund, by the Canada Council for the Arts, by the Ontario Arts Council, by the Federation for the Humanities and Social Sciences through the Awards to Scholarly Publications Program, and by the University of Ottawa.

Copy editing: Michael Waldin
Proofreading: Thierry Black
Typesetting: Édiscript enr.
Cover design: Édiscript enr.

Library and Archives Canada Cataloguing in Publication

Therapeutic touch: research, practice and ethics / edited by Martin Rovers, Judith Malette and Manal Guirguis-Younger.

Includes bibliographical references and index.
Issued in print and electronic formats.
ISBN 978-0-7766-2547-8 (softcover).
ISBN 978-0-7766-2548-5 (PDF).
ISBN 978-0-7766-2549-2 (EPUB)

1. Touch—Therapeutic use. I. Malette, Judith, 1963-, editor II. Guirguis-Younger, Manal, 1967-, editor III. Rovers, Martin, 1949-, editor

RC489.T69T54 2017 615.8′22 C2017-904809-0
C2017-904810-4

Printed and bound in Great Britain
by Marston Book Services Limited, Oxfordshire

Canada Council for the Arts Conseil des arts du Canada

ONTARIO ARTS COUNCIL
CONSEIL DES ARTS DE L'ONTARIO
an Ontario government agency
un organisme du gouvernement de l'Ontario

Canada

Table of Contents

CONCLUSION

Acknowledgements

My (Martin) personal sincere thanks go to Manal and Judith for their support, wisdom, and academic ability in making this book possible. It is one thing to have a few good ideas, but it is a blessing when academic peers come together to make it a reality. This wonderful book would never have been be possible without them.

To my peers at Saint Paul University who have supported my work in couple and family therapy over these years, and allowed me to venture into the realm of touch in counselling and the helping relationships. To my long-lasting friend, Philip Domingue, who introduced me to Stan Tatkin's writing many years ago, which led to our conference at which Stan was our speaker in 2014. To Cassandra Petrella and Michael Machen, two students now well graduated from our Saint Paul University program in couple and family therapy, who were the first to help me write about this field of inter-partner touch in couple counselling.

To Tom O'Connor, my colleague in the Society for Pastoral Care and Counselling Research (SPCR), for his continuous support and humour over the 20 years we have been a part of SPCR, and who has graciously accepted to write the introduction to this book. A special thanks goes to SPCR and the John Wells Foundation for their generous publication grant for this book as well as the conference on touch in 2014 from which this book sprang.

To all the authors who have worked hard to put their ideas and research on paper and make this edited book real. This book is rich in the variety of theoretical approaches and chapters on touch in the helping relationships, and in the many places and ways touch is a part of the healing process.

To Jane Chambers, MA, for editing and coordinating the book and making it ready for the publisher.

We three editors wish to express appreciation to Saint Paul University for their publishing grant for this book, for their ongoing support for our research, conferences, and writing, and especially to the professors and students in the Faculty of Human Sciences, where touch, as a dimension of counselling and spiritual care, is taught and practiced every day.

Martin Rovers, PhD
Judith Malette, PhD
Manal Guirguis-Younger, PhD

Introduction

Exploring Touch

Thomas St. James O'Connor

"To touch or not to touch, that is the question." This question is an adaptation of Hamlet's famous soliloquy in William Shakespeare's play (Shakespeare, 1950, Act III, Scene 1, l.56f; also see Zur and Nordmarken, 2009). If Hamlet were a psychotherapist today rather than the Prince of Denmark in the Middle Ages, he would be faced with this existential and therapeutic dilemma. This book on touch, edited by Martin Rovers, Judith Malette and Manual Guirguis-Younger (2017), explores that question and invites readers to investigate the use of proper and caring touch in the context of psychotherapy.

The authors offer a variety of perspectives on the topic of touch. The image of a garden in full bloom is an apt way to describe this book. In a garden, there is often a rich variety of plants, colours, textures, odours, tastes and sounds, which is a simile for the multi-dimensional aspect of touch. The sense of touch is rich and complex. Touch is woven into the other senses, part of a whole. We often use words for touch in relation to the other senses. For example, I was "touched by what someone said." That is a common phrase and shows the link between hearing and touch. The sound waves touch my ear drum, and the words, emotion and insight of the speaker impact (touch) my spirit. Light waves touch my eyes so that I can see. This book is an invitation to explore the richness of the garden and to discover what it contains. Yes, some forms of touch are abusive, the forbidden fruit in the practice of psychotherapy; however, do not let the fear of forbidden fruit stop you from exploring the positive aspects of this garden. Adam and Eve in the Hebrew story walked with God in the garden and enjoyed God's touch, and indeed touched God (Gen. 2: 20–25; note that all Scripture passages are from The Jerusalem Bible, 1967). That is the beauty, enchantment, and blessing of the garden (Moore, 1996). There was only one tree that they were forbidden to touch; God wanted them to enjoy the garden. There are many rich surprises in this garden, this book, and please touch the plants.

As each chapter notes, touch is a subject that is not sufficiently addressed in education or in psychotherapy training. In my own education and training, I was told not to touch clients; except for shaking hands, any other form of touching was a boundary violation. However, even shaking hands has come into question with an awareness of cultural diversity; for some cultures, shaking hands is not acceptable, especially between genders. Yet, the issue of touch in counselling has been present for a long time. We are not angels but human beings with a physical body and a sense of touch. The question that Hamlet, the psychotherapist, asks is not new.

One approach is to remove touch completely from counselling, that is, to stay out of the garden because of fear of touching the forbidden fruit. Psychotherapy by phone, email, and video conferencing does that. While these forms of therapy are growing and have a place, very few psychotherapists would argue that these should replace face-to-face encounters between therapist and client. There is a growing body of research that indicates that caring touch is necessary for human functioning and for humans to thrive, from birth to death. To live in a world where there is the absence of caring touch or very limited touch can produce physical death as well as emotional and spiritual death. This book addresses this controversial topic with academic rigour.

Touch does come into play in almost every helping relationship. There is continual touch between a nurse and the patient for whom care is given. Touch often takes place in therapy and supervision through handshakes. A distressed client might receive a touch on the shoulder from the therapist in a moment of pain during a session (see Patricia Berendsen's essay in Chapter V). One partner might reach out and hold the other partner's hand at the invitation of the therapist (see Cassandra Petrella and Martin Rovers in Chapter VII). Many persons touch their dogs and pets when they are distressed. Pets can often sense the distress of the owner and will reach out and make contact (see Kristine Lund in Chapter VI).

Touch requires contact with the other. When I stroke my dog, Lucky, his fur touches my hand as well as my hand touches his fur. He can move away or he can remain. To touch is to be touched; however, the other does not always give consent. In abusive touch, whether physical or sexual, the other does not consent and in fact is wounded by the touch. The experience of abuse through touch is often what therapists work hard to heal in psychotherapy. Many clients come for help from psychotherapists because of the wounds from abusive touch and/or the absence of touch. The sense of touch is not a simple phenomenon. It can be interpreted in many different ways; our interpretations of touch are greatly influenced by family and cultural values.

This book addresses many different aspects and contexts of touch, including supervision, ethics, palliative care, counselling in marriage and family therapy, and others. It is multi-voiced and written by a group of diverse scholars and clinicians, exploring the topic and exposing the complexity of the issue. If someone is looking for a simple answer to the issue of touch in therapy, to Hamlet's question, they will be disappointed. I invite you to enjoy the complexity and see it as strength and enjoy this walk through the garden.

The exploration of the question is further enriched by the variety of standpoints of the authors. These authors' expertise draws on various disciplines: nursing, philosophy, biblical studies, cultural anthropology, marriage and family therapy, psychology, Islamic studies, spiritual care (chaplaincy), psychotherapy, social work, ethics, and more. They explore the strengths and limitations of touch within their respective practices. This multidisciplinary approach is refreshing. Each author is aware of the dangers and misuses of touch. They enter into the discussion with caution, fear and trembling, and a sense of exploration and enchantment.

Overall, the various writers encourage touch in their clinical work, but with some reservations. Each has re-entered the garden. They are aware of the negative consequences of improper touch. In almost every chapter, some of the dangers and the ethical issues around touch are exposed. Again and again, education, training and supervision are promoted and required for caring touch in psychotherapy. Most importantly, client consent is necessary. Each author also emphasizes caring touch and how important caring touch is to human well-being.

Caring touch ought to be used appropriately between therapist and client, religious leaders and religious followers, and between partners and family members. The book invites the reader to explore caring touch in the practice of helping and in human relationships in general.

The complexity and richness of touch include religious and spiritual perspectives. As noted above, touch, for good and evil, is present in the Hebrew and Christian Scriptures. Chapter IV considers a biblical understanding of the touch that Jesus used. His healing power was often expressed through touching the distressed person, the unclean and abandoned person, the "untouchables." People reached out and touched Jesus to experience healing; however, there were times that he did not touch nor wanted to be touched. Chapter VIII, written from an Islamic perspective by an Islamic scholar and spiritual care provider, examines the laying on of hands, which is used by imams and others as a traditional practice of healing in Islam. Chapter IX reflects on spiritual practices around sacred objects. This qualitative research explores how participants touch sacred objects that they often carry with them, and how this touch offers comfort and hope in stressful times. Some of the chapters include spiritual aspects to touch. Spirituality in therapy is a topic that arose over 20 years ago. Prior to that, spirituality and religion were not subjects often raised by psychotherapists with their clients. It was then considered forbidden fruit, much like touch is today. However, the research by Harold Koenig and colleagues (2001), Ken Pargament (2007), Augustus Meier and colleagues (2004), and many others indicate that spirituality and religion are important supports for many people and should be examined and discussed in psychotherapy. Spirituality and touch are closely connected, similar to the body's connection to the soul.

Research is a necessary part of the exploration of touch in psychotherapy. The College of Registered Psychotherapists of Ontario (CRPO) makes research and reading professional literature one of the five competencies that psychotherapists are required to have in the practice of psychotherapy (CRPO, 2012). There are many methods of research, and this book utilizes a number of them. The dominant method is hermeneutical in the investigation of touch. Hermeneutical research examines and interprets texts and ideas. Other methods are also used. For example, Chapter VI employs a mixed method of a quantitative survey and qualitative means in a study of touching and seeing dogs in answering the research question: Can dogs lower the stress among university students at exam time? Chapter IX presents qualitative research and a summary of a Master's thesis on participants' experiences of using sacred objects in dealing with stress. In many of the chapters, there are case studies, both short and long, and they illustrate and put flesh on the theories advanced by the authors on the use of touch. Case illustrations are a great help to clinicians, and a form of research. With the dominant method being hermeneutical and the inclusion of other methods such as quantitative, qualitative, and case study, this book makes a contribution to the field. Academics and clinicians as well as general readers can profit from reading this book. The various forms of research offer insight into an evidence-based practice on touch in psychotherapy. While the use of touch with clients is at the beginning stages in terms of research, some initial findings have emerged. The various chapters provide many kinds of evidence for this practice of touch. As one who has worked in a medical setting for a few years, I can attest to the need for evidence-based health care for effective patient care.

The first four chapters provide an excellent theoretical base in the areas of counselling, philosophy, cultural anthropology, and theology. They offer different theories and perspectives on touch. The other chapters, as well, present some theory of touch related to the topic of the chapter. Certainly, more theory could be added to this large field on touch. The aspect of gender and touch is only briefly mentioned. Cultural variations and specifics on touch need more attention (Zur, 2007). In general, more theory and open conversation by academics and psychotherapists is needed on touch. The case studies also show a strong clinical base. This book is a great resource to both academics and clinicians. All the authors advocate that clinicians who venture into touch should do so with caution, self-reflection, and supervision. Since this is an area for potential harm to clients, the caution is necessary and wise. The conclusion draws together the many and complex themes in the book. The conclusion offers a brief summary of the themes on touch and psychotherapy that emerge in the various chapters, holding the tension between benefits and harm of touch in psychotherapy. If Hamlet the psychotherapist were to continue his soliloquy, it might go like this:

> To touch or not to touch, that is the question:
> Whether it is more noble and ethical to sit in my therapist chair
> And remain aloof by not reaching out to touch a client in need
> Or to take a risk and touch the shoulder of a client

who is in pain, awash in a sea of troubles.
This action could be the death of my practice
If understood as inappropriate
Resulting in disciplinary action and/or litigation.
Do I have client consent?
But if I am not willing to risk failure and censor for the good of
 a client,
How can I serve clients? What kind of therapist am I?
I believe I touch the client for his/her own good—
or is this a dream?
Have I misled myself? And does touch comes from my need?
I need more education and supervision.
To touch or not to touch, that is the question
Or maybe the question is,
What is appropriate touch, and how do I talk to clients about this?

Enjoy your walk in this rich garden of touch.

Outline of the Book

Part I deals with the theory of touch and explores the religious and philosophical dimensions of touch. Chapter I begins with "Touch Deprivation and Counselling as Healing Touch," by Isaac Davis, Martin Rovers, and Cassandra Petrella. These authors draw on research to show the importance of caring and healing touch in human relationships and therapeutic relationships. Deprivation of touch can be very harmful, especially to infants, just as abusive touch is harmful.

In Chapter II, Richard Feist provides a philosophical approach to touch in "Thinking About Touch." Feist describes several views by world-renowned philosophers on their understanding of touch. He begins with Plato and moves to contemporary times. Chapter III, by Christian Bellehumeur and Jane Chambers, uses cultural and sensory anthropological interpretations in "Contributions of Sensory Anthropology and Durand's Anthropology to the Symbolic Study of Touch and the Understanding of Boundaries in Psychotherapy." Drawing on the writings of anthropologist Gilbert Durand and the notions of boundaries in psychotherapy, the authors argue that touch varies from culture to culture, and within cultures. The cultural context of the client and therapist are important points to remember in dealing with touch in psychotherapy. Finally, this section on theory ends with Karlijn Demasure's chapter on "Healing and the Forbidden Touch: A Reflection on Selected Scripture Stories (Mk 5: 25–34) and Mary Magdalene at the scene of the Resurrection (Jn 20: 11–18)." This chapter analyzes these two biblical passages concerning Jesus and touch. In the first passage, Jesus is touched by a woman and she is healed. He is aware of the touch and does not condemn her for that. In the second passage, Mary Magdalene approaches Jesus in the garden after the resurrection and he asks her not to touch him. Demasure compares and contrasts these two different responses of Jesus to touch.

Part II addresses the various practices and contexts of touch. In Chapter V, Patricia Berendsen discusses touch between therapist and client in "The Intervention of Touch in Psychotherapy and Trauma Treatment." Based on theory and clinical practice, she argues that touch can be helpful for clients with trauma, under certain conditions. A case example illustrates her rationale. Chapter VI, by Kristine Lund, focuses on research examining the effect of touching and seeing puppies by university students around exam time. In this chapter, "A Puppy's Touch: De-stressing with Dogs in a University Environment," Lund found that participants lowered their stress levels from a brief encounter with puppies. Drawing on other research as well as her own research study, she argues that touching a friendly dog can lower stress. Cassandra Petrella and Martin Rovers write about inter-partner touch in couple counselling in Chapter VII. These two authors present "Inter-Partner Touch in Couple Counselling: Theory and Emerging Practice," and argue that couples who are struggling in their relationship can build a more secure base by appropriate touch through eye contact, hand holding, and 20-second hugs. They draw on the theory of John Bowlby, Sue Johnson, and Stan Tatkin in presenting their argument. They also use a case study to illustrate the theory.

Part II also includes touch in spiritual and palliative care, and in nursing and supervision. The Muslim laying of hands used by imams is described in Chapter VIII. Nazila Isgandarova, an Islamic scholar and spiritual care provider, describes this practice in "The Practice of the Laying on of Hands in Islamic Spiritual Care." She presents the strengths and weaknesses of this approach and notes that it is not used in health care institutions. The Muslim laying on of hands is mostly practiced by the imam in the mosque. Linda Mayorga Miller describes a qualitative study that she did, in Chapter IX. Her chapter, "An Angel in My Pocket: Touch, Sacred Objects and Spiritual Coping," presents the results from her qualitative study in which she examines participants' experience and use of sacred objects to help in coping with stress and adverse situations. She found that these sacred objects, carried and touched by participants. are as valuable as prayer and meditation in coping. In Chapter X, "Touch(ed) in Palliative Care Nursing: Moving with/in an Uncertain Practice," two nurses, Lacie White and Christine McPherson, describe the use of touch by nurses. The authors present three stories in which touch is used. They outline some of the dangers and limitations of touch in nursing; however, they argue that touch can provide comfort and support to people who are dying. Touch can also help family members. Cynthia Bilodeau ends Part II with Chapter XI, where she describes supervision theory and practice around touch in "Touch in Supervision." She notes that the issue of touch between supervisor and supervisees is not part of the literature. Supervisors have been wary of touching supervisees out of ethical and boundary concerns; supervisors often urge supervisees not to touch their clients. There has been nothing written on supervising caring touch between a supervisee and the client.

The final part, "Ethics of Touch and Conclusion," has but one chapter and the Conclusion. In Chapter XII, Marilyn Guindon, Reesa Packard, and Natalie Charron present an extensive review of the literature on the ethics of touch in psychotherapy. Their contribution on "The Ethics of Therapeutic Touch" cautions practitioners and academics against a naive understanding of touch as helpful. The authors warn about the potential abuse, but they do not come up with any definite answer on whether to touch or not to touch. Informed consent by the client is necessary, and appropriate caring touch that is helpful to the client is also needed. The Conclusion synthesizes the previous 12 chapters and offers an integrative overview. Martin Rovers, Judith Malette, and Manal Guirguis-Younger present the closing remarks in "Issues of Touch: An Overall View and Integration." These authors note that touch helps us thrive, and yet abusive touch and the deprivation of touch produce the opposite. They also point out the difficulty of discussing this issue with clients, as it might raise defensiveness and insecurity in the client; however, consent by the client is required for touch in psychotherapy.

References

Jerusalem Bible. (1967). New York: Doubleday.

College of Registered Psychotherapists of Ontario (CRPO, 2012). Research Competencies. Retrieved from http://www.crpo.ca/wp-content/uploads/2013/06/RP-Competency-Profile.pdf

Koenig, H., McCullough, M. and Larson, D. (2001). *Handbook of religion and health*. New York: Oxford University Press.

Meier, A., O'Connor, T. and VanKatwyk, P. (Eds.). (2005). *Spirituality & health: Multidisciplinary explorations*. Waterloo, ON: Wilfrid Laurier University Press.

Moore, T. (1996). *Re-enchantment of every day life*. New York: HarperCollins.

Pargament, K. (2007). *Spiritually integrated psychotherapy*. New York: Guilford Press.

Shakespeare, W. [1598] (1950). *Hamlet*: *Complete works of William Shakespeare*. Vol. 3, Toronto: Leland.

Zur, O. (2007). *Boundaries in psychotherapy: Ethical and clinical explorations*. Washington, DC: American Psychological Association – APA Books.

Zur, O. and Nordmarken, N. (2009). To touch or not to touch: Exploring the myth of prohibition on touch in psychotherapy and counseling, clinical and ethical. Retrieved from www.zurinstitute.com/touchintherapy.html

PART I

THE THEORY OF TOUCH

Chapter I

Touch Deprivation and Counselling as Healing Touch

Isaac Davis, Martin Rovers and Cassandra Petrella

For children to form a secure attachment style to their main caregiver figure, the caregiver must be not only emotionally available but also physically available in a timely, consistent, and appropriate manner (Bowlby, 1969; Ainsworth, 1989). Children require consistent love, attention, and touch from their caregivers. They need to experience warm loving touch, to be hugged, held, and cuddled. Basic childhood needs are primarily met through touch. For the child, this leads to feelings of safety and love (Bowlby, 1988; Clulow, 2001; Gillath, Selcuk and Shaver, 2008). However, if the child is deprived of loving touch, what message does the child take on about attachment and security within relationships? If there is a lack of touch between caregiver and child, what might the effects be? How is attachment wounded? Sadly, the earliest studies and research on attachment and touch come from studies on touch deprivation (Bakwin, 1942; Spitz, 1945; Harlow, 1958; Carlson et al., 1995). In brief, touch deprivation over the first years of life negatively influences the infant's relational ability throughout life. A review of this literature on touch deprivation enables counsellors to grasp the significance and power of touch in the creation of safe attachments. Implications for counselling will be outlined and two case studies included. This chapter opens with a brief look at how we see touch today so we can better understand touch deprivation.

What Is Touch?

One of the most important elements of humankind is the ability to develop and maintain secure attachments and meaningful relationships with one another. Human relationships are paramount to human survival as mammals; infants need their caregiver to survive. Furthermore, secure attachments and touch are needed to learn, to work, to love, and to procreate. Secure attachments take many forms; however, the most influential, wide-ranging, pleasurable and, at times, painful attachment relationships are the ones developed with intimate loved ones—friends, family, lovers. Inside this circle of intimate relationships, humans are bonded together through love (Perry, 2001).

From the moment one is born, the brain begins to form and develop the ability to attach and to love (Hofer, 2005). Through acts of holding, cuddling, feeding, gazing, kissing and many other forms of nurturing behaviours, caregivers/parents begin to send the message of love and safety to the child through touch. This bond of love will continue to grow and develop through quality time, eye contact, physical proximity, smell, sound, and taste. The number one way infants give and receive communication is through touch (Field, 2001). Thus, in developing a secure attachment between child and parent, touch is arguably the most important element in the early stages of development.

Our entire body is covered with this amazing sense system, our skin, fully equipped and capable of receiving information about the surrounding environment and communicating it to our brain (Field, 2001). Our skin is the first sense organ to develop within the womb (Field, 2001; Montagu, 1971). It is incredible that the fetus, while in the womb suspended in amniotic fluid, begins to experience the sensation known as touch (Field, 2010). After birth, the newborn begins to receive significant amounts of stimulation through cuddling, breast feeding and eye contact. It is this contact comfort (Harlow, 1958) that is essential for growth and development (Perry, 2001), as well as calming infants in pain or discomfort (Field, 2010). Furthermore, research has demonstrated "kangaroo care," skin-to-skin contact between parent and infant, which sends positive messages to the infant and aids the infant in being able to regulate their own bodily state (Ferber, Feldman and Makhoul, 2008). Undoubtedly, touch is the most developed sense at birth, and touch has been found to significantly contribute to the cognitive, brain and socio-emotional development in infancy and childhood (Field, 2001).

Touch Deprivation

Early understanding of the benefits of touch comes from studies and research on the negative effects of the absence of touch. Touch deprivation is defined as a lack of tactile stimulation of our largest sensory organ, the skin, particularly in early infancy (Field, 2001). Touch deprivation in infancy would be the lack of cuddling, holding, hugging, eye contact, cooing, and the like. As touch is needed for sufficient attachment development, when it is not present or removed during infancy and childhood, long-lasting negative effects take place, in terms of attachment injuries and touch wounds.

Infants and children who have been raised in or spent significant time in institutional care—for example, in hospitals, orphanages or residential care—typically receive minimal touch from caregivers (Field, 2001). In most of these institutions, caregivers are caring for several children at one time, making it almost impossible for the caregiver to spend the adequate amount of time necessary to provide the amount of physical stimulation that the child requires in developing secure lifelong attachment ability.

Lowrey (1940) conducted studies on the ill effects of hospitalization of children. He had observed children developing a variety of behaviours, such as hostile aggression, shyness, difficulties eating food, stubbornness, and excessive crying, to name a few. Lowrey concluded that children should not be raised in institutions (hospitals) because children experience an isolation-type effect due to the lack of community and touch. It is because of this isolation that the children were beginning to develop maladaptive behaviours (Van der Horst and Van der Veer, 2008).

Even more influential was the work of Bakwin (1942) who studied young children being cared for at New York's Bellevue Hospital. The hospital had been experiencing a very high mortality rate among the young children living there. Initially, the cause of death was believed to be malnutrition and the spread of infection throughout the hospital. In order to avoid infections spreading further among the children, the open ward concept of the hospital was replaced by small cubicle rooms where each child had their own room, isolated from everyone else. The only people allowed into the children's cubicles were nurses and physicians, wearing hooded and masked attire. Even the children's parents were not allowed to enter the cubicles (Bakwin, 1942). In order to stave off malnutrition, children were put on a high caloric diet. However, even with these measures in place, their mortality rate did not decline. Furthermore, the children only started to gain weight when they returned home.

Bakwin (1942) believed the sterile environment was not helpful but was actually damaging the children because of the total lack of touch. He understood that children need to receive tactile stimulation and remain in close proximity to the caregiver in order to survive. His conclusions led to a drastic change in hospital policy. Nurses were permitted and encouraged to "mother" and "cuddle" the children, to touch them, to pick them up and play with them. Furthermore, parents were invited to visit with their child, which previously was not allowed (Van der Horst and Van der Veer, 2008). The results were dramatic. Even though the risk of spreading infection was dramatically increased, the mortality rate for infants under one year of age fell substantially, from 30–35% to less than 10% (Bakwin, 1942). Bakwin concluded that, in infancy, the loneliness element involved in being separated from other people and the lack of touch may not only be undesirable, but fatal to infants (Van der Horst and Van der Veer, 2008).

Spitz (1945) compared infants living with their mothers in a prison nursery with infants living in a foundling home, a hospital for abandoned children. Spitz claimed the infants in the foundling home started to deteriorate after they were weaned; it was at this point where contact with any person was reduced to essential only levels. On the other hand, children in the prison nursery system appeared to be thriving.

Spitz did acknowledge that the circumstances in the foundling home were not ideal. For instance, the infants in the foundling home had minimal toys and were never moved from their cots; thus the infants could only stare at the ceiling and, most importantly, there was only one nurse tending to several infants. The infants were not being held, cuddled, or hugged by the nurses, and therefore, they did not receive sufficient amounts of tactile stimulation. Outside of feeding and changing, the infants were mostly deprived of human touch. On the other hand, the infants in the penal nursery were being taken care of by their mothers and by other women. Spitz described these mothers as lavishing their love on their babies. Spitz concluded that separation from loving touch caused the foundling infants to suffer significant damage to their well-being (Spitz, 1945, as cited by Van Rosmalen, 2012).

Spitz later conducted a follow-up study on the infants. He found the children in the foundling home were experiencing disastrous long-term effects. Of the original sample, approximately one-third of them had already died. The children that remained in the foundling home were severely delayed physically and mentally (Spitz, 1945, as cited by Van Rosmalen et al., 2012). Although they were between the ages of two and four, only half of them were able to spoon-feed themselves. In this study, only one could formulate whole sentences, and all of the children were undersized for their age. Spitz concluded that irreparable emotional and attachment damage happened during the infants' first year of life, even when their basic physical needs of food, water, and shelter were met. Conversely, the children in the penal nursery were very active, playing games, interacting with one another, and overall doing well. Further research led Spitz to conclude that separation from caregiver/mother, that is, separation from the infant's love object for a prolonged amount of time, should be avoided at all costs. However, if unavoidable, care should be taken to provide a sufficient caretaker substitute for the child to attach to (Spitz, as cited by Van Rosmalen et al., 2012).

Harry Harlow conducted several experiments with rhesus monkeys (Harlow, 1958; Harlow, 1959; Harlow and Zimmerman, 1958). He had previously lost a batch of monkeys to disease and was determined to ensure his new batch of infant monkeys would remain in excellent health. Therefore, for reasons of hygiene and nutrition the monkeys were separated from their mothers at birth. By separating the monkeys from their mothers, Harlow was unknowingly setting up an environment very similar to Bakwin's hospital and Spitz's foundling home.

The monkeys in his study, who did not receive "contact comfort," developed and exhibited strange behaviours, such as sucking their fingers and toes and rocking back and forth (Harlow, 1958). Furthermore, some developed brain damage; others were violent and socially impaired as adults. Additionally, Harlow found that when a monkey mother was unable to provide contact comfort to her newborn infant, it could, more often than not, be traced back to a lack of bonding between the mother and infant. Thus, when infant monkeys do not receive the necessary contact comfort from their mothers, they are challenged to learn safe bonding and comfort. This leads to an inability to provide it to their own offspring later in life. Touch deprivation, and similarly, touch, is learned and contagious. Harlow developed the term "contact comfort," which relates to the need to feel touch and security between mother and child based on the findings of his study.

Children adopted from Romanian orphanages were found to be delayed in all areas of development due to a lack of contact comfort (Chisholm, Carter, Ames and Morrison, 1995). Results have further found that the lack of touch, the main form of communicating love to infants, leads to cognitive delays (Maclean, 2003) and neurodevelopmental delays (Chugani et al., 2001; MacLean, 2003). When compared to same-age children, children who have been deprived of touch and a significant caregiver score below average in cognitive skills. Even after adoption, it appears the developmental delays persist for several years (Beckett et al., 2006).

Touch deprivation does not only exist in institutions. It also occurs for infants whose caregiver/mothers suffer from depression (Field, 2001). One study found that infants of depressed mothers spent greater quantities of time touching themselves as compared to infants of non-depressed mothers. This was believed to compensate for the lack of touch the infant was receiving from their mother. These infants were also found to be more aggressive or vigorous in their touching, grabbing, patting and pulling for the things they desired (Field, 2010).

Through these studies it is imperative to realize that tending to the basic needs of children, such as providing food, water, shelter, and clothes, is not sufficient in promoting healthy development. Separation from family, a consistently loving caregiver, or not receiving physical stimulation or touch was found to be extremely detrimental to the child's health and development (Van der Horst and Van der Veer, 2008).

Touch Deprivation and Attachment

The previous studies outlined touch deprivation in terms of the failure to meet a child's physical and emotional needs. This section will explore touch deprivation and attachment in particular. Attachment styles fall into two categories: secure and insecure; and under the insecure, two main styles are noted: avoidant and preoccupied (Bowlby, 1969; Ainsworth, 1989; Rovers, 2005). Main and Solomon (1990) added the term "disorganized" to describe patterns of attachment observed in the "Strange Situation" experiment, for when a child appeared to lack an organized strategy in regards to the attachment figure.

Children raised in institutions (orphanages, for example) are at an increased risk of experiencing a variety of social and behaviour problems, such as developing an insecure attachment style (Zeanah, 2000). A study by Tizard and Rees (1975) identified a group of 65 children who had been placed in a London residential nursery at birth or shortly thereafter; 24 of the children were adopted between the ages of two and four, another 15 were returned to their birth families. The remaining 26 stayed in the institution. The 26 children remaining at the institution were assessed at age four. Only eight children had been able to develop some form of attachment to a caregiver. Eight children showed no signs of discriminated attachment but displayed characteristics of an avoidant style due to being emotionally withdrawn and unresponsive. As well, 10 children were put into an indiscriminate category: These children showed no preference between a stranger or family member but would simply approach and seek attention from whomever was available (Tizard and Rees, 1975).

There have been two longitudinal studies following children adopted out of Romanian institutions. Indiscriminate behaviour, that is, showing no preference between strangers or family but seeking attention, was found to be one of the most relevant social abnormalities in children adopted out of institutions (Chisholm, Carter, Ames and Morrison, 1995; Chisholm, 1998). Security of attachment, as reported by parents, significantly increased in these children, between 11 and 39 months, post-adoption. However, for children who were adopted after eight months of age, their level of indiscriminate friendliness did not diminish; the majority of these children would still approach strangers for comfort (Chisholm, 1998). A study by O'Connor, Marvin, Rutter, Olrick, Britner and ERA Study Team (2003) found that 73% of the highly indiscriminate children would be classified with an insecure disorganized attachment style.

Vorria, Papaligoura, Dunn, Van IJzendoorn, Steele and Kontopoulou (2003) investigated attachment style by using the Strange Situation for children at a Greek institution. Within the institution there were books and toys available to the infants; unfortunately, the caregivers did not have enough time to play with the infants and the materials made available. The caregivers only had enough time to feed, bathe and clothe the infants. What's more, infants spent approximately 18 hours a day in their cots. Results found that 66% of the children studied were classified with a disorganized attachment with their caregivers; 24% were considered to be securely attached and the remaining

10% were unclassifiable (Vorria et al., 2003). Compared to a group of infants reared at home, 25% had a disorganized attachment style and 41% were secure. This illustrates the power of the surrounding environment and the need for children to be touched (held, cuddled, hugged, and so on) and stimulated in order to feel safe enough to grow and develop (Vorria et al., 2003).

Another study utilized the Strange Situation procedure for young children in a Romanian institution, assessing the children's attachment to their "favourite" caregiver. The study found that only 18.9% of children had developed a secure attachment, 3.2% avoidant, 65.3% disorganized, and 12.6% unclassifiable (Zeanah, Smyke, Koga and Carlson, 2005). In contrast to the never-institutionalized group, 74% were classified as secure, 4% avoidant, and 22% disorganized. The institutionalized children were in a chronic state of touch deprivation, receiving minimal quality contact or love from anyone (Zeanah et al., 2005).

Disorganized attachment is associated with the infant's exposure to prolonged or repeated separation from the caregiver, maltreatment, maternal depression, and high marital conflict (Vorria et al., 2003). As previously stated, disorganized attachment has presented as very high in institutional care for infants as well as other high-risk situations. However, approximately 15% of infants raised in homes with apparently normal functioning caregiver/mothers were still classified disorganized (Van IJzendorn, Schunegel, and Bakermans-Kranenburg, 1999). An argument has been put forth by Main and Hesse (1990) that disorganized attachment tends to develop in infants who have experienced their parents as frightening. Therefore, when the child experiences distress, they are confused about approaching the caregiver for a resolution of the anxiety-provoking event because the caregiver is a source of both comfort and fear and anxiety (Vorria et al., 2003). Disorganized attachment appears to be prevalent for children raised in institutional care due to the fact that the care is, more often than not, discontinuous, fragmented, lacking intimate touch and unable to meet the individual attachment needs of each child (Lyons-Ruth and Jacobvitz, 1999).

Touch, or the lack thereof, always plays a role in the attachment style developed in children in all families. The Strange Situation experiments (Ainsworth, 1989) found that attachment styles in children can be classified according to the caregiving style of the caregiver/mother involved in the child's life. Securely attached children tend to have caregivers who were consistently available, avoidant attached children tend to have caregivers who were consistently unavailable, and preoccupied children tend to have caregivers who were inconsistently available.

Counselling as Healing Touch

Working with individuals and couples who have experienced touch deprivation in childhood creates a difficult challenge for counsellors. We have seen from the above studies that, without sufficient touch, infants suffer socially, cognitively and physically, and are at high risk of developing an insecure attachment style. Levels of touch and consistency of childhood attachment do not determine an individual's life outcome; however, it certainly can create attachment injuries and wounds for later life, and impact the quality of their interactions in intimate relationships.

Counselling is about helping relationships heal. Counsellors work with people who have suffered attachment wounds from childhood experiences of touch deprivation or attachment inconsistency. The counsellor's task is to teach clients new ways to soothe attachment injuries and promote a more secure attachment stance with the self and significant others in life. Some of these ways of healing touch can be taught, as in cognitive behavioral therapy, while other healing ways can be caught, as in creating new experiences that enable touch and attachment to be a safe experience again. Counsellors need knowledge of the neurobiological aspects of attachment as well as sensitivities in the practice of healing touch to create comfort in connection again. The pages that follow attempt to outline some basic approaches to healing touch in individual and couple counselling; some of these approaches will be expanded on elsewhere in this book.

It is very difficult to change one's attachment style, but it does not dictate the rest of one's life. Perry (2001) described learning new attachment as similar to learning a new language. If one learns a language at a young enough age, one will speak the language without an accent. However, if one learns the language later in life one will still be able to speak the language but will do so with an accent. When learning or changing one's attachment later in life, the individual will not feel as comfortable with the attachment as if they had learned it in infancy; they will have an accent but, more importantly, they will be able to speak the language of increasingly secure attachment.

Touch has been shown to be a healing factor across cultures and history (Stockham-Ronollo and Poulsen, 2012) and physical affection has been linked to relationship and partner satisfaction (Gulledge and Fischer-Lokou, 2003). Given that the human brain retains its plasticity much longer than initially believed, the brain can still learn, grow and adapt to new experiences and information of touch. Cognitive behavioural therapy teaches us that change is possible through repetitive cognitive restructuring of behaviours and thoughts. Basic cognitive behavioural principles reinforce that the more we touch, the more we will become accustomed to touch, just like any behaviour. For example, the client can be asked to reach out to others more through touch in everyday life. Cognitively, the client's fears about touch can be reality-tested to discover if these fears are truly founded in real risk or if they are only the residue of childhood attachment wounds. Talk therapy addresses the cognitive process of the brain, but in order to experience long-lasting growth the brain needs to change at multiple levels (Lucas, 2012).

Self-touch can be a soothing coping strategy as well for clients experiencing symptoms of anxiety or depression. Based upon concepts of Somatic Experiencing therapy (see Chapter V), clients are guided to connect with their present issues through self-soothing practices. In one somatic exercise, the therapist directs the client to close her/his eyes, find a position that feels most comfortable to them, and to wrap their arms around themselves in a hugging embrace. Then the client is asked to pause on the posture that feels most right. The client is given a few moments to just sit and feel that soothing position. Once the client opens her/his eyes the therapist will ask the client if she/he has any guesses as to why this position feels most comforting. Often, the client will respond that her/his caregiver held her/him that way. Perhaps the client's mother once held her/his hand that way, and so the position mimicked that handholding touch. It is comforting to the client because it is the same way the client's mother reached for them when she/he was crying as a child. The felt sense of that handholding has stayed with her/him. The client would then be prompted to start holding her/his hand in that way in the future when feeling upset about something. The somatic position exercise can prove soothing during difficult times and can be used as a helpful coping technique for clients to psychobiologically soothe themselves outside of therapy when they experience distress.

Somatic Experiencing therapy found that touch can be an important and powerful intervention for trauma treatment (Berendsen,

2014). Touch that is sensitively utilized by a skilled and well-trained counsellor can assist in self-regulation, an absent characteristic of traumatized clients. Adhering to safety principles and acquiring knowledge about nervous system regulation, as well as the threat response cycle, supports the foundation for integrating touch into counselling.

In focusing therapy, a mindfulness-based approach, individuals are often coached towards a calm and meditative state in which they are directed to find the bodily sense of what it is that is troubling them (Berendsen, 2014). With their eyes closed and their body relaxed, they are asked to place their hand on an area in their body where they feel a twinge of distress. For example, clients with anxiety may report they feel a lump in their throat at the thought of confronting their coworker about an error they blamed on them. The therapist instructs clients to keep their hand there and describe that feeling, as clients make sense of the part of them that feels distress because they fear they are not worthy of having a voice; they begin to process that fear. Then as the therapist guides clients to ask that part of them what it is that it needs, clients complete part of the healing for this wound of theirs. The therapist uses the client's own sense of touch to connect with an emotional issue through a psychobiologically mindfulness-based approach.

Canine-assisted therapy (see Chapter VI) found positive change in how clients were feeling as a result of interacting with and touching puppies. Interacting with an animal can be an important resource to manage stress. This resource is of particular importance to clients who have had previous positive experiences with animals. Providing animal-assisted counselling and activities may be a low-cost resource to support clients and promote a greater sense of well-being.

By combining talk therapy with a variety of other approaches, counsellors will be better able to meet the needs of the individual and couple clients who come for healing of insufficient or deprived touch in childhood. Touch can be an integral part of clinical interventions when appropriate. There is hope for relearning touch through counselling as more and more empirically based therapies work on the neurobiological level to create change within individuals.

Healing Touch in Couple Counselling

If inter-partner touch is utilized effectively it can allow partners to meet emotional and physical needs, repair the emotional bond between partners who are under distress, and can strengthen anxious partners' connection to each other. Several couple counselling approaches have begun to integrate touch into therapy.

Solomon and Tatkin (2011) have developed a psychobiological approach (PACT) to couple therapy. This approach utilizes interpartner touch during a couples counselling session. Techniques used in PACT during therapy include eye-to-eye contact, holding hands, moving toward and away from each other, and direct embracing. These exercises are designed to aid the couple in developing a sense of comfort with proximity with their partner and foster touch and connection between them. Solomon and Tatkin (2011) believe that gentle skin-to-skin contact has a powerful effect on people as the desire to be touched, held and cuddled in a safe, nourishing environment is hardwired into the human brain (Lucas, 2012; Solomon and Tatkin, 2011; Tatkin, 2011). Touch, our most basic human contact between child and caretaker, is a reminder of childhood. Touching in adult relationships can, oftentimes, bring up strong emotional issues. This corresponds with Montagu (1971) when he asserted that touch is many times stronger than verbal contact, and it affects everything we do. No other sense can arouse us like touch. Solomon and Tatkin (2011) believe that being able to learn to better interact through touch and proximity, and develop skills to be able to regulate one's own and one's partner's feelings and emotions, is essential in regulating the emotional distress that can arise within a couple as old attachment and touch wounds surface. Partners become the new safe touch environment for each other. The psychobiological approach to couple therapy is explored more by Petrella and Rovers in Chapter VII of the book.

In emotion-focused couple therapy (EFT), Sue Johnson believes that touch is one of the more basic bonding mechanisms. There is minimal physical contact in EFT; however, Johnson will regularly ask the couple to look at each other in the eye and communicate to their partner what they are feeling. Johnson (2008) states that, without regular touch, it is very difficult, if not impossible, for a couple to create a safe, intimate environment where emotional bonding takes place. Even sex without daily touch appears to lack bonding. Johnson argues it is touch—the hugging, holding, caressing, and stroking that evoke the emotions—which, she states, are the royal route to love.

Case Study 1

Perry (2006) shares his story of a teenage mother, Virginia, and her four-year-old daughter, Laura. Their story demonstrates how children respond to early experience, how the brain works, and how the brain has the ability to recover.

Virginia had brought Laura to the hospital because she needed help with her struggling daughter. Laura was severely underweight, weighing merely 26 pounds at the age of four. Upon arrival at the hospital, doctors had decided to put Laura on a high calorie diet. However, even on this high calorie diet, Laura continued losing weight. Doctors were perplexed at how this was possible; some even suggested that she must be purging when no one was looking.

Virginia's background was troubling. As a child, she was raised through the foster care system. She was abandoned by a drug-addicted mother and her father was unknown. Furthermore, Virginia grew up in a time when the child welfare system did not want children to become too attached to any one family or caregiver; therefore, she was moved every six months to a new home. At that time the importance of developing a secure attachment figure had not penetrated the child welfare bureaucracy.

When Virginia gave birth, she had no idea how to tend to the emotional needs of her baby, as we cannot give what we have not received. She would feed her child sitting up, not nuzzled close and safely to her body. There was minimal touch, minimal contact comfort between them. Virginia only knew how to meet the basic needs of her child: feed, clothe, bathe and sleep. She didn't rock, cuddle, or coo with her daughter, or stare into her eyes. Laura was being deprived of touch, of contact comfort; the emotional signal of love and safety was non-existent. Research has demonstrated that in order for mammals to grow, physical stimulation (touch) is required for the body to feel safe; only when it feels will the body release growth hormones (Field, 2001; Perry, 2006). Without these physical and emotional signals, Laura's body began to shut down, refusing to release growth hormones.

To become a parent capable of creating a safe, loving environment for her daughter, Virginia had to learn it. To do this, Virginia and Laura went to live with a lady nicknamed Mama P. who had extensive experience working with children who had been victims of abuse and neglect. She believed these children needed to receive physical stimulation in the form of cuddling, hugging, being held, rocked, and comforted, even if this sort of physical stimulation appeared more appropriate for younger children. She held the belief that all children need this type of physical contact. If children had not received it in their younger, more sensitive years, then they would need to receive it in later years. Children who were not loved and touched as infants and young children are still starving for touch, as if their skin is hungry for touch (Perry, 2006).

During the first month living with Mama P., Laura consumed the same number of calories that she had been receiving in the hospital. As previously stated, at the age of four, she weighed 26 pounds. In Laura's new loving, nurturing, and physically stimulating environment, she gained an astonishing 10 pounds in one month. Her body weight increased 35% on the same number of calories. The power of touch communicated to her the emotions of love and safety, equipping her brain to release the much-needed and appropriate hormones required for growth (Perry, 2006).

Case Study 2

Tiffany has been attending counselling individually with the author (Isaac) for several weeks. While working with Tiffany, I learned she had been raised in a very dysfunctional home. Her father was an alcoholic and verbally abusive. At the age of 14, her father left the family, leaving them with no money. Tiffany recalls days where they had no food; she was totally ashamed of the vehicle they had to drive, as the door had to be tied on with bungee cords. Shortly after her father left, Tiffany's mother had a mental breakdown. At this point, Tiffany assumed the role of head of the household, protector and caregiver.

Tiffany remembers constantly feeling in danger, recalling frightening incidents, such as their furnace breaking down and causing a fire, her bike being stolen, and her house being broken into. Tiffany was constantly on alert, always afraid something bad was about to happen to them. However, through all of this, Tiffany recalls that her greatest pain and greatest wound was, in fact, her mother. Tiffany never knew what she was going to get with her mother. She would be happy and sweet one minute and the very next she would be yelling and screaming. She never knew if it was safe to approach her or who to go to in times of need. Thus, Tiffany learned to keep her head down, to take care of herself, and stay out of people's way. This could be seen before and in session, as I never heard her enter the waiting room: She was incredibly quiet, not wanting to disturb anyone who might scold her. At work she would tiptoe around and would consistently hold her breath, as it was not even safe to breathe because it would make too much noise.

Currently, Tiffany is happily married to a good man, but she is scared at work and home, and she struggles with extreme anxiety and catastrophic thoughts. She struggles to regulate her own body, thoughts and emotions. Talk therapy was effective for her, but more was needed to help her connect more fully with her husband.

Tiffany knew her husband was a good man, but she didn't know how to feel safe and calm around him. Because she trusted her husband, we were able to introduce an exercise called the 30-second hug. Simply put, her goal was to hug her husband for 30 seconds. This is a non-sexual hug: it is simply to be held and rest in each other's arms. This task may seem simple to some but it is very daunting for many. At first, as expected, she struggled to relax and be calm so close to a caregiver. However, slowly and persistently, she found herself becoming more relaxed. Tiffany reported that simply listening to his heartbeat, his deep breaths in and out, began to help her regulate her own heartbeat and breathing, as if they were becoming in tune with one another.

Eventually, she was able to speak to him about what made her feel more comfortable during the hug: She found he held on a little too tight sometimes, which was his own insecurity coming out, as he was afraid she would push him away. They were able to talk about the hug, to help each other better meet their needs for contact; she was able to reassure him she would not run anywhere and that he did not need to squeeze her too tightly.

This hug gave them an opportunity to connect, for them to experience safety and comfort with each other. She did not experience full safety and comfort right away; it took some practice and communication on how to improve in order to make her feel more comfortable. Slowly, over time and with practice, she began to feel more secure and comfortable in his presence.

Conclusion

The first few years of a child's life is a period of rapid brain development; early childhood experiences have a disproportionate impact on the shaping of our neural systems, shaping the way we think and our attachment style. Children raised in institutions are at higher risk of developing an insecure attachment style due to the lack of contact with a primary caregiver. However, even children raised in regular family homes can experience touch deprivation. It is extremely difficult for any one parent to meet every need of a child, and all parents fall short in some way. An insecure attachment style does not mean you are broken, but it can cause pain, confusion and unhappiness, and creates difficulties in future relationships.

People are not lost causes; they simply have work to do in order to feel safe and secure in relationship. The good news is they can do it. Some strategies are developed in childhood to survive: to cope with the emotional pain of not getting touched enough, not receiving eye contact, and crying and having no one respond; avoidance is a way to cope with this pain. The strategies in childhood are not useful in adulthood. Nobody had perfect parents; attachment wounds are not our fault, but as adults, it is our responsibility to work through our attachment wounds towards new and more secure relationships. Touch can get us there because it has the ability to bond in ways that words cannot. When we feel cerebrally connected to our partner, this felt sense can calm us in ways that will open our hearts to more love and connection, which is the ultimate wish for us all.

References

Ainsworth, M. D. (1989). *Patterns of attachment: A psychological study of the strange situation*. New York: Eribaum.

Bakwin, H. (1942). Loneliness in infants. *American Journal of Diseases in Children, 63*, 30–40.

Beckett, C., Maughan, B., Rutter, M., Castle, J., Colvert, E. and Groothues, C. (2006). Do the effects of early severe deprivation on cognition persist into early adolescence? Findings from the English and Romanian adoptees study. *Child Development, 77*, 696–711.

Berendsen, P. (2014). Supporting the integration of the body into psychotherapy and trauma treatment. In T. O'Connor, K. Lund and P. Berendsen (Eds.), *Psychotherapy: Cure of the soul*. Waterloo: Waterloo Lutheran Seminary.

Bowlby, J. (1969). *Attachment and loss, Vol. 1: Attachment*. New York: Basic Books.

Bowlby, J. (1988). *A secure base: Parent-child attachment and healthy human development*. New York: Basic Books.

Carlson, M., Dragomir, C., Earls, F., Farrell, M., Macovei, O., Nystrom, P. et al. (1995). Effects of social deprivation on cortisol regulation in institutionalized Romanian infants. *Society for Neuroscience Abstracts, 21*, 524.

Chisholm, K. (1998). A three-year follow-up of attachment and indiscriminate friendliness in children adopted from Romanian orphanages. *Child Development, 69*, 1092–1106.

Chisholm, K., Carter, M., Ames, E. W. and Morison, S. J. (1995). Attachment security and indiscriminately friendly behavior in children adopted from Romanian orphanages. *Development and Psychopathology, 7*, 283–294.

Chugani, H. T., Behen, M. E., Muzik, O., Juhasz, C., Nagy, F., and Chugani, D. C. (2001). Local brain functional activity following early deprivation: A study of post institutionalized Romanian orphans. *Neuroimage, 14*, 1290–1301.

Clulow, C.F. (2001). *Adult attachment and couple psychotherapy: The "secure base" in practice and research.* London: Brunner-Routledge.

Ferber, S. G., Feldman, R., and Makhoul, I. R. (2008). The development of maternal touch across the first year of life. *Early Human Development, 84*, 363–370.

Field, T. (2001). *Touch.* Cambridge, MA: MIT Press.

Field, T. (2010). Touch for socioemotional and physical well-being: A review. *Developmental Review, 30*, 367–383.

Gillath, O., Selcuk, E. and Shaver, P. R. (2008). Moving toward a secure attachment style: Can repeated security priming help? *Social and Personality Psychology Compass*, 2(4), 165–1666.

Gulledge, N. and Fischer-Lokou, J. (2003). Another evaluation of touch and helping behavior. *Psychological Reports, 92*, 62–64.

Harlow, H. (1958). The nature of love. *American Psychologist, 13*, 673–685.

Harlow, H. (1959). Love in infant monkeys. *Scientific American*, 200(6): 68–74.

Harlow, H. and Zimmermann, R. (1958). The development of affectional responses in infant monkeys. *Proceedings of the American Philosophical Society, 102*(5): 501–509

Hofer, M. A. (2005). The psychobiology of early attachment. *Clinical Directions in Neuroscience Research*, 4(4–6), 291–300.

Johnson, S. (2008). *Hold me tight: Seven conversations for a lifetime of love.* New York: LittleBrown.

Lowrey, L. (1940). Personality distortion and early institutional care. *American Journal of Orthopsychiatry, 10*(3), 576–585.

Lucas, M. (2012). *Rewire your brain for love.* New York: Hay House, Inc.

Lyons-Ruth, K. and Jacobvitz, D. (1999). Attachment disorganization: Unresolved loss, relational violence, and lapses in behavioural and attentional strategies. In J. Cassidy and P. R. Shaver (Eds.), *Handbook of attachment: Theory, research and clinical applications* (pp. 520–554). New York: Guilford

MacLean, K. (2003). The impact of institutionalization on child development. *Development and Psychopathology, 15,* 853–884.

Main, M. and Hesse, E. (1990). Parents' unresolved traumatic experiences are related to infant disorganized attachment status: Is frightened and/or frightening behavior the linking mechanism? In M. T. Greenberg, D. Cicchetti and E. M. Cummings (Eds.), *Attachment in the preschool years* (pp. 161–181). Chicago: University of Chicago Press.

Main, M. and Solomon, J. (1990). Procedures for identifying disorganized/disoriented infants during the Ainsworth Strange Situation. In M. Greenberg, D. Cicchetti and M. Cummings (Eds.), *Attachment in the preschool years* (pp. 121–160). Chicago: University of Chicago Press.

Montagu, A. (1971). *Touching the human significance of the skin.* (2nd ed.). New York: Harper and Row, Publishers.

O'Connor, T. G., Marvin, R. S., Rutter, M., Olrick, J. T., Britner, P. A. and the English and Romanian Adoptees (ERA) Study Team. (2003). Child-parent attachment following severe early institutional deprivation. *Development and Psychopathology, 15,* 19–38.

Perry, B. (2001). *Bonding and attachment in maltreated children consequences of emotional neglect in childhood.* New York: W. W. Norton and Company.

Perry, B. (2006). Skin hunger. In *The boy who was raised like a dog and other stories from a child psychiatrist's notebook.* New York: Basic Books.

Rovers, M. (2005). *Healing the wounds in couple relationships.* Toronto: Novalis Publications.

Solomon, M. and Tatkin, S. (2011). *Love and war in intimate relationships: Connection, disconnection, and mutual regulation in couple therapy.* New York: W. W. Norton.

Spitz, R. A. (1945). Hospitalism: An inquiry into the genesis of psychiatric conditions in early childhood. *The Psychoanalytic Study of the Child, 1,* 53–74.

Stockham-Ronollo, S. and Poulsen, S. (2012). Couple therapy and reiki: A holistic therapeutic integration. *The Family Journal: Counselling and Therapy for Couples and Families, 20*(3), 292–298.

Tatkin, S. (2011). *Wired for love.* Oakland: New Harbinger Publications, Inc.

Tizard, B. and Rees, J. (1975). The effect of early institutional rearing on the behavior problems and affectional relationships of four-year-old children. *Journal of Child Psychology, Psychiatry, and Allied Disciplines, 27,* 61–73.

Van der Horst, F. and Van der Veer, R. (2008). Loneliness in infancy: Harry Harlow, John Bowlby and issues of separation. *Integr Psych Behav,* 2008 (42), 325–335.

Van IJzendoorn, M. H., Schuengel, C. and Bakermans-Kranenburg, M. J. (1999). Disorganized attachment in early childhood: Meta-analysis of precursors, concomitants and sequelae. *Development and Psychopathology, 11,* 225–249.

Van Rosmalen, L., Van der Horst, F. and Van der Veer, R. (2012). Of monkeys and men: Spitz and Harlow on the consequences of maternal deprivation. *Attachment and Human Development, 14*(4), 425–437.

Vorria, P., Papaligoura, Z., Dunn, J., van IJzendoorn, M. H., Steele, H. and Kontopoulou, A. (2003). Early experiences and attachment relationships of Greek infants raised in residential group care. *Journal of Child Psychology and Psychiatry, 44,* 1208–1220.

Zeanah, C. H. (2000). Disturbances of attachment in young children adopted from institutions. *Journal of Developmental and Behavioral Pediatrics, 21,* 230–236.

Zeanah, C., Smyke, A., Koga, S. and Carlson, E. (2005). Attachment in institutionalized and community children in Romania. *Child Development, 76*(5), 1015–1028.

Chapter II

Thinking About Touch

Richard Feist

The title "Thinking About Touch" might seem to be about two things that are as far apart from one another as any two things could be. After all, thinking is often regarded not only as a very abstract activity in and of itself, but perhaps the most delicate of all ways in which the mind can reach out. The image of one engaged in thinking is often of a mind shut off from the world, a mind that is blocked from the world but enjoys a radical liberty nonetheless. Moreover, this activity of thinking, free of the world, is able to grasp things that we cannot via our hands. For example, the mind can think about the difference between an apple's colour and shape, the number seven, and whether or not the Good is identical to the Beautiful. Touch is arguably the most concrete way in which the organism contacts its world. Touch is anything but an abstraction; it is perhaps of all the senses the most deeply plugged in to the corporeal world. Touch seems to require the body, contact with objects, and physical motion—in short, pretty well everything that thought does not.

These preliminary observations do not square well with the history of philosophy. For example, Aristotle linked thought and touch in a deep manner. Human flesh, he wrote, was the softest of any creature, and the most sensitive. Through flesh we have touch, and as Aristotle held, touch is common to all life. This extreme sensitivity of touch that we humans have, Aristotle thought, is why we are the most intelligent of all creatures (Aristotle, 1984, p. 421a). Aristotle even associates touch with the highest form of interaction in the cosmos: knowledge. "In participating with the intelligible, the intellect thinks of itself, for it becomes intelligible by touching and thinking" (Aristotle, 1984, p. 1072b20-1). In this chapter, some reflections on touch are offered, a sense that has been, in the history of philosophy, at times valued and at times ignored. The main goal is to motivate the view that touch is in fact quite difficult to pin down in terms of its necessary and sufficient conditions, and that the examination of touch shows that the mind's experience of the body is a rich and diverse commodity, one that is not handed ready-made, but requires interpretation.

Hume on Thinking

To bring thought and touch together, let us take some guidance from David Hume. According to Hume, all our thoughts are made up of ideas. These ideas, Hume insisted, were never innate; instead, they are derived from experience (Hume, 1978, p. 1–5). Hume calls the mind's contents "perceptions." He then divides perceptions into two main categories: impressions and ideas. Hume writes:

> The difference betwixt these consists in the degrees of force and liveliness, with which they strike upon the mind, and make their way into our thought or consciousness. Those perceptions, which enter with most force and violence, we may name *impressions*; and under this name I comprehend all our sensations, passions and emotions . . . By *ideas* I mean the faint images of these in thinking and reasoning . . . (p. 1)

Impressions are produced when the body interacts with the world. These impressions can be either simple or complex. A simple impression might be of the colour blue, and a complex one is when you look, say, at a cityscape from the top of a skyscraper. Ideas are directly related to impressions. See the colour blue, get the impression of it, and then close your eyes and recall the colour blue. This memory of blue is basically an idea of blue. Clearly, if you have not seen blue before, then you will not be able to have an idea of it. This simple idea of blue is a copy of the simple impression of blue. The relationship between complex ideas and complex impressions is, not surprisingly, more complicated than that between simple ideas and simple impressions. Your idea of Paris is not a copy of your impression of Paris.

Moreover, you have ideas, such as those of unicorns and pixies, of which you have no corresponding impressions. Hume's view is that the idea of a unicorn is, in the end, simply an amalgamation of the idea of a horse and the idea of a horn.

There is an important link with touch when it comes to thought, according to Hume. How does one tell the difference between eating an apple and simply imagining eating an apple? This is simple, provided that one helps oneself to the presupposition that the imagined act of eating the apple is only occurring in the mind and that the actual act of eating the apple is occurring in the real world. Hence, the difference is simply that one is imaginary and one is real. Suppose, though, that one does not have this presupposition. Indeed, ever since the time of Descartes, the existence of the external world has been a philosophical problem. It is not a problem in the sense that any serious philosopher truly does not believe in the external world's existence or even seriously doubts it. Rather, the problem has been how to justify the claim that there is an external world and not simply assert it and hold to it on faith.

If we put aside the unjustified presupposition that there is an external world, then how does one distinguish between the imagined apple-eating and the real apple-eating? Hume's answer is simple: by their feel. As we have seen, impressions have a certain intensity to them, a vivacity, as Hume says. Eat a piece of chocolate and then recall eating the chocolate. There is no difference in the actual content of the impression and the memory, but nobody would prefer the memory of eating chocolate over actually eating chocolate. The idea is the impression drained of nearly all vivacity, which makes it a memory; but drain the impression of all vivacity and one then has a pure idea. When we move these ideas around in our minds, then, we are thinking. Thought, in other words, is very much like the trace of touch. Thought as the trace of touch is an old theme in philosophy. One can find it in Hobbes and as far back as Saint Augustine. This discussion does not enter into the thorny question of whether thought as the trace of touch only says that thought content is the trace of touch and leaves open the notion of thought processes, namely rationality, and their relation to touch.

This view of Hume's makes the thought picture somewhat too abstract, placing it up and inside the mind, as though after a while it occurs independently of the external world. Hume certainly did not hold this notion of a radically independent subject; instead, he thought that the subject was, in many ways, a fiction created as a coping mechanism by an intrinsically pathological mind. I will return briefly to the role of the subject when I discuss the unisensory view of touch that later authors held.

Now I will turn to Descartes, who is often regarded as articulating a strong mind-body dualism, but in fact offers us a picture that is more likely closer to the truth, albeit it serves to problematize the whole notion of touch and its relationship to thought, even more so than that of Hume. I will leave aside Descartes's complicated view of innate ideas and describe his view of thought and sensation.

Descartes: Touch as Confused Thinking

In the opening sections of his masterwork *The Meditations,* Descartes investigates the nature of his self, striving to find what is essential to it, and what is not. I am not going to rehearse Descartes's investigation of the self here, but suffice it to say that through hyperbolic doubt, such as the famous Demon arguments and arguments from imagination, Descartes concludes that the essential aspect of the self is that it thinks. Anything about the self can be rationally, that is consistently, doubted away, but to doubt that the self thinks is, in the end, an act of thinking. In one of the most famous passages from the *Second Meditation,* Descartes poses the question, "What am I?" and answers with "a thinking thing (*res cogitans*)." What exactly is this *res cogitans?* Descartes responds with not so much of a description of what a *res cogitans* is, but what a *res cogitans* does, a performative description, if you will.

"A thing which thinks. What is that? A thing which doubts, understands, affirms, denies, wills, is unwilling, and also imagines and feels" (Descartes, 2000, p. 110). The curious and rather startling aspect here is that Descartes includes willing, imagining and feeling as examples of thinking, a very wide extension of thinking to say the least. Many Cartesian scholars have argued that such an extension was commonplace when Descartes was writing (Anscombe and Geach, 1971, p. XXXVII), but this is contestable. The medieval thinkers, such as Aquinas, separated the mind into several faculties, and so willing and thinking and sensing were fundamentally different activities, not simply modes of one activity. Nonetheless, even if willing, imagining and so on are modes of thinking, that does not mean that the modes are identical to one another. In two other passages Descartes clarifies the invariant in these modes. The first is in his second reply to the critics of *The Meditations*.

> By the word "thought" I include everything that is in us in such a way that we are immediately aware of it. Thus all the operations of the will, understanding and imagination, and senses are thoughts. But I added "immediately" to exclude those things that follow from these operations, such as voluntary motion, which surely has thought as its principle but nevertheless is not itself a thought. (p. 162)

Clearly then, Descartes is not saying that the nature of a particular content renders that content a mode of thought; instead, it is that content's relationship to the mind that renders it a mode of thought. In this case, the relationship is one of immediate awareness. For example, when one wills X, senses Y, imagines Z, one is directly aware of X or Y or Z. This direct awareness is incorporated into the very act of willing, sensing and imagining.

There are two possibilities of the object of awareness here; that is, is the mind directly aware of the content of the act or of the act itself? Descartes never clarified this in the above texts, but as John Cottingham pointed out, Descartes did provide some illumination in a recorded conversation with Frans Burman, held near to Descartes's death (Cottingham, 1978, p. 211). Descartes explains to Burman that "being aware" is a reflective act of the mind. In other words, the manner in which the act is seen by the mind, immediately, is what makes that act a thought.

This appears to lead Descartes into trouble since it seems to make everything going on into a thought and slides into identifying the modes. There is a key discussion in *The Meditations* that helps illustrate how Descartes ultimately interpreted the notion of sensations, which brings us back to the central issue of this paper, the relationship between thought and touch. Descartes tells us that imagining and sensing are special modes of thinking. That is, I can understand myself without them but I cannot understand them without presupposing myself. In other words, one can understand a *res cogitans* that only has rational thoughts (but no imagination and no sensing), but one cannot understand the very act of imagining or sensing without also including some notion of understanding (Descartes, 2000, p. 135). Not only do sensing and imagining have this non-symmetric aspect to them, but there is something still more about them, which has to do with their content. From here, I will concentrate solely on Descartes's view of sensing.

Descartes writes that it is via the senses, the sensations of pain, hunger, thirst and so on, that I clearly see that I am not in my body in the way "a sailor is present in a ship." If a ship is damaged, the sailor perceives that damage in a disinterested manner. I do not perceive my bodily damage in a disinterested manner. Hence, I am not a pure *res cogitans* in a body; rather, ". . . I am most tightly joined and, so to speak, commingled with it, so much that I and my body constitute one single thing" (Descartes, 2000, p. 136). This discussion is not simply limited to bodily damage, but is Descartes's reflections on the whole spectrum of sensing: taste, touch, sight, smell, and sound. Descartes concludes:

> For clearly these sensations of thirst, hunger, pain, and so on are nothing but certain confused modes of thinking arising from the union and, as it were, commingling of the mind with the body. (Descartes, 2000, p. 136)

In another example, Descartes compares our situation with an embodied angel. An angel is a pure *res cogitans*, and were it embodied, it would not feel its body as we do ours. Rather, it would perceive only motions caused by external objects (Cottingham, 1978, p. 213). The angel would be embodied as a sailor in a ship.

In sum, Descartes points out that we are not a pure *res cogitans* that happens to be embodied. Instead, we are a *res cogitans* inexplicably and inextricably intertwined with a body. Much of our intellectual life is pure thought; we can in some cases think like an angel, but in many cases, such as our sensory experiences, our thought is confused perceptions. These sensations, these confused perceptions, are composed of a reflective part (the awareness) and the content. Cottingham describes the content of confused perceptions nicely, as a ". . . curious residual element, which might be called the 'qualitative feel' (e.g., the painfulness or hurtiness [*sic*] of pain), that remains, in terms of Descartes' metaphysics, ultimately mysterious" (Cottingham, 1978, p. 213).

If we translate this discussion into terms of touch, then we have something like the following. To describe the touch of say, a soft pillow, as "I feel softness" would have the clarity of the awareness of feeling something soft, but the actual softness felt is not clearly and distinctly known. It is a brute given in thought, not something that can be rationally known, only felt. It is a place where thought and touch are closest, but the latter remains rationally impenetrable. Descartes's views on thought and sensation varied over his lifetime. My discussion here was meant to capture the essence of his mature thought. For Descartes's intellectual development, see Gaukroger (1995).

What Is Touch?

From Hume and Descartes we see that thought and touch are indeed closely related. If the previous discussion is plausible, then one would expect that the attempts to rationally ascertain just what touch is would have difficulties. Indeed this is precisely the case. Numerous attempts have been made to deal with touch; one of the classic papers is by Grice (1966). Grice offers four basic characteristics of touch, which have been developed in various directions. One such development is that of Nudds (2003). Nudds offers an overall description and analysis of the "counting question"—simply put, how to count the senses, how to differentiate them. One upshot of Nudds's discussion is that there are presuppositions made in counting, namely, how to differentiate the domain upon which the counting will occur. Counting is not simply reading off things in the world directly.

The issue of ascertaining just what touch is, though, requires some clarification. In this paper, my concern is to ask the following: What is touch and how is it demarcated from the other senses? The goal is to see whether one can truly speak of what criteria uniquely and completely designate touch. My view is that there is not such a set of criteria. In other words, the grand picture that I hold is that touch, along with the other senses, is highly interactive. Indeed, one can separate them, but only if one employs numerous conventions. Otherwise, there is no fact to the matter of what uniquely and completely designates touch.

Experience, then, is a mélange of our senses, and how we carve them up is much like how we carve constellations out of the pattern of stars. Clearly the stars and their spatio-relationships are there, but exactly what pictures we make out of them is conventional. This line of argument is much along the greater lines that all of nature, from that information coming from the outside to the information coming from the inside, must be interpreted. Nature inside or nature outside does not come neatly cut for us; we must cut it up ourselves. This does not mean that just anything goes in this carving of nature. For the ways that nature is carved must not violate laws of logic and should, as the pragmatists liked to say, have good cash value. All this does mean that the carving should not be read, as Plato might say, a cutting of nature at its joints.

Clearly, I am not going to argue for this entire conventionalist picture here; but suffice it to say that touch is there but not as a natural kind: intrinsically demarcated as though it could exist independently. To help illustrate this point, I wish to offer some further reflections on touch.

Individuating Touch

The question of individuating the senses, that is, specifying the criteria that are complete and unique to a specific sense, is an old philosophical chore. When conducting an examination of the foundations of knowledge, Plato, in his *Republic,* has Socrates point out that the powers of the soul are a "special class of things" that enable us to do the things that we do. Sight, for instance, is a soul-power, as is hearing. Socrates continues:

> A power has neither color nor shape nor any feature of the sort that many other things have and that I use to distinguish those things from one another. In the case of a power, I use only what it is set over and what it does, and by reference to these I call each the power it is: What is set over the same things and does the same I call the same power; what is set over something different and does something different I call a different one. (Plato, 1997, pp. 477c–478b)

There are a number of important points in this passage. First, note that a power, that is, the sense in question, lacks distinguishing predicates. For instance, we may use a predicate like blue, to distinguish between two things, such as a blue chair and a non-blue chair. However, a power, that is, a sense itself, like sight or hearing, cannot be blue, or not blue. So to distinguish the special set of enabling powers of the soul, we have to use what is set over against those powers. To distinguish sight from hearing, we use visual objects and auditory objects, those things that are set over against the power, not any mark of the power itself. Plato then moves on from the sensorial powers of the soul to its intellectual powers.

Perhaps the earliest, sustained study of the individuation of the senses was conducted by Aristotle in *De Anima*, Book II, Chapter 6. Aristotle awarded a high place to the notion of touch. The organ of touch, ultimately, was located near the heart, near the location of where Aristotle thought our intelligence rested. This raises an interesting question as to how Aristotle would have conducted a phenomenology of the body (which he did not), since the feeling of the center would have been in the chest and not, as we have it, behind the eyes. This is a topic for another time. Aristotle praises touch as follows:

> While in respect of all the other senses we fall below many species of animals, in respect of touch we far excel all other species in exactness of discrimination. That is why man is the most intelligent of all animals. (Aristotle, 1984, p. 4201a20–23)

Aristotle picks up on Plato's approach that the senses are defined and clarified via an examination of their objects. Again, like Plato, Aristotle's concern with the individuation of the senses is within a larger project, namely, the investigation of the nature of the soul. To properly investigate the soul requires that one delineate the powers of the soul: thinking, perceiving, nutrition absorption and so on. Of course, in order to delineate and articulate each of these powers, one needs to first describe the act of perceiving and, in turn, one must describe that which is being perceived, namely, the object of the perception: in this case, the object of the sense in question. However, Aristotle goes further than Plato in his pursuit of the demarcations of the senses. Not only does a demarcation of a sense refer to that sense's object but also the physiological process involved (Aristotle, 1984, p. 403a25-b9). Aristotle did not, of course, have access to our modern understanding of physiological processes; for instance, his view of the physiological process involved in sight was that there was a discoloration of the eye-jelly. Nonetheless, the description of the physiological process itself involves reference to the sense organ, in this case the eye, which in turn takes on (or participates in) a characteristic of the sense object. That is, the eye-jelly becomes coloured in the way that the seen object is coloured.

In his study of Aristotle's program of sense individuation in *De Anima*, Richard Sorabji argues quite rightly that we must make an important disclaimer on Aristotle's behalf, that is, that Aristotle is not claiming that reference to sense objects will provide a necessary and sufficient condition for the demarcation of the senses. Instead, Aristotle is

> interested in demarcating the sensory powers that a zoologist can expect to find in this world, rather than ones which could exist in logically possible worlds. So in assessing the criteria by which he demarcates the senses, we ought to consider whether they are adequate for this zoological purpose. (Sorabji, 1971, p. 59)

In sum, this coheres with what we have seen with Descartes. The senses are clearly there, that is, they are powers that we have; however, their internal markings are not usable in a conceptual framework. Descartes thought that their content was immediately present to the mind, but confused, not amenable to a conceptual framework or theorizing. Aristotle, following Plato, holds that their content can be somewhat articulated, although he does not say exactly how. In the end, the ultimate manner of individuation involves a reference to the sense object itself and that the task of individuating the senses falls within the contours of a general organizational project in zoology, and not a philosopher's attempt at the articulation of a given sense's necessary and sufficient conditions.

The issue of individuating the senses, especially with regard to touch, is still an ongoing debate within philosophy. The key thing is that to individuate a sense X with respect to X's sense object, Y, is inherently circular. If I am looking for the necessary and sufficient conditions of touch, for instance, and I then claim to have the correlates of this sense, that is, the touchable objects, which I collect in a set, Q, the problem is to be able to have legitimate confidence in Q. If I am not sure or do not know the criteria for X, how can I be sure that I have the right members for Q? How did I go about choosing the members for Q in the first place? How would I even recognize them? (This is, without doubt, a version of the Meno problem, long made famous by Plato's dialogue of that name.) Moreover, I could be charged with reading an illegitimate ontology into the soul based on the presupposition that the mind's powers are divided along the lines of the world's objects. An even broader question is that of the initial selection of X. How did I select X as a separate portion of experience in the first place? Also, what about the impact of the sciences on the accepted (or acceptable) definition of touch? For instance, some have argued for the existence of therapeutic touch, which incorporates ideas from quantum mechanics and offers a very different notion of touch, namely, as the interplay between electromagnetic fields generated by human bodies. For the last section of this paper, I wish to consider a few more contemporary thinkers on the problem of the senses, again with a concentration on touch.

Twentieth- and Twenty-First-Century Touch

Sorabji, in his discussion of Aristotle, points out other criticisms of the attempt to individuate the senses via the objects of the senses. He concludes that such an approach, while helpful, ultimately does not do the full job of demarcation (Sorabji, 1971, p. 67). Turning specifically to touch, Sorabji notes that touch is the sense that has an overwhelming number of possible objects (Sorabji, 1971, p. 68). This view of touch's wide range is well documented in the philosophical literature and is often referred to as the "heterogeneity problem" (Fulkerson, Unity, 2011; Ratcliffe, 2012).

Aristotle, then, in recognition of the heterogeneity problem, moves away from the object-based approach to individuating the senses when it comes to touch. His predecessors, having recognized the special and difficult status of touch, offered two criteria for its individuation: contact and non-localizability. Aristotle opts for contact. His reasoning is quite straightforward since it is entailed by his metaphysics. To fully explicate this metaphysical entailment is beyond the needs of this paper; for my purposes, it suffices to note that for Aristotle, an object, such as a chair, is composed of two irreducible, basic elements: matter and form. A chair is composed of matter, say, wood, and the form is what makes that particular amount of wood into a chair. A living organism is also composed of two basic elements, matter and soul. Matter is the physical aspect of the body and it is the soul is (like a form) that makes that matter into a living body. For Aristotle, all of the senses, except touch, are the body's way of capturing forms from objects. For instance, to see a chair is not to come in any kind of contact with the chair; no matter from the chair has any kind of impact upon the eye. According to Aristotle, a copy of the chair-form "moves" from the chair to the eye. So if contact is needed, in the sense of matter-matter contact, only touch will require it. Hence, touch can be circumscribed via contact. My point is not one of textual exegesis, but of philosophical understanding. In this case, the issue is one of defining terms: What do we mean by contact? Must it be defined as matter-matter?

Clearly, Aristotle and his contemporaries meant "physical contact." Other thinkers, such as Democritus, who subscribed only to a physical universe, held that all senses have to be in some kind of contact with their objects. Hence, for Democritus, talking about touch and characterizing it by contact would not serve to individuate it. Aristotle, who embraces non-physical forms in his philosophy, could then say that those senses which deal solely in the appropriation of forms do not have any contact with their objects, and so contact will serve to individuate touch. One might think that such an argument would not stand up today given that modern science embraces physicalism; there are no forms of any kind in physics textbooks. But this does not mean that contact is any more easily analyzed. Moreover, there are numerous difficulties regarding the notion of contact; such difficulties are often debated in metaphysics and mathematics (Hazen, 1990). The same kind of ontological play could be said to be at work in some modern thinkers' approaches, although they do not embrace Aristotle's particular view of matter and form. Some thinkers today have argued that touch might not always include contact since it is possible to perceive something with which we have no contact (Fulkerson, Touch, 2012; Ratcliffe, 2012).

These non-contact arguments turn on the manner of characterizing that which we are experiencing. For example, suppose that we are listening to a note played on a piano, say middle C. The physical story is that the piano's mechanism manages to vibrate the air at such a frequency and that these vibration patterns enter the ear canal, and it is those vibration patterns that we have contact with, but what we perceive is middle C, not the vibration patterns. So, the argument goes, we are perceiving middle C, we only have contact with the air vibrations, and since middle C is not air vibrations, we are perceiving something with which we have no contact. It is tempting to reply that such arguments are muddles of confusions, but a more charitable and useful approach is to note what such arguments do: They make a number of presuppositions. In this case, the presupposition is that middle C (as heard) is different from the associated air vibration, but what exactly does this presupposition mean? Certainly, an air vibration is not identical to an experience of middle C, since an air vibration is just a pattern of movement in the atmosphere, whereas the experience of middle C is a complex of ear, mind, interpretation, and vibration patterns in the atmosphere. On the other hand, it is not self-evident that middle C is some kind of object that we experience here and with which we have no contact. The vibration pattern is part of the constitution of the experience of middle C. How does one show that middle C is radically separate from the pattern? The philosophers who push these kinds of arguments do not, in my view, answer this question. Instead, the arguments ultimately assert the difference and separation between the proximal stimulus, the air vibration, and the hearing of middle C. Again, I would maintain that all of this, from air vibration to hearing of middle C, is a large, connected complex. On my reading, then, contact is germane to all of the senses and so cannot be used to isolate or individuate any of them. My reading, of course, could be challenged, but the point is that even the notion of what constitutes contact is not self-evident and must be provided with a context. The upshot of this discussion is that trying to lay down a criterion for touch, such as contact, is not a simple matter, for criteria do not drop like manna. Criteria themselves have contexts, that is, presuppositions, and hence are contestable.

Similar kinds of presuppositions are made when philosophers argue for a weaker position regarding the senses. In the case of touch, a weaker view is that although we may not be able to come up with some kind of necessary and sufficient condition that uniquely identifies touch, we can at least establish that touch is a unique sense, or in the current terminology of the philosophical literature, that touch is unisensory and not multisensory (Fulkerson, Unity, 2011). In general, we move from the strong individuating position, that the senses are unique and that we can provide an articulation of each one's uniqueness, to the weaker view, that the senses are unique, but we cannot exactly say why or how. The motivation for this weaker view lies in the position that arguments for unique criteria fail but there are good arguments for unisensory accounts. This weaker thesis resembles what Wittgenstein used to call "family resemblance concepts" or what are sometimes deemed "cluster concepts" (Wittgenstein, 1958, p. 32). For example, I may easily tell the difference between the Jones family and the Smith family, but how do I do this? A strong thesis, for example, would be via necessary and sufficient conditions: that there is a criterion (or criteria) that is (are) necessary and sufficient to being one of the Joneses and not one of the Smiths. This might be something like a property, Z, that all and only the Joneses have. As Wittgenstein noticed, families rarely, if ever, obey such a simple and strong rule. Instead, there are a variety of things, a particular nose, eyebrows, hair colour and so on, that run throughout a family. No single property is instantiated by every member of the family, but the family is nonetheless a kind of unity and so can be distinguished uniquely and so separated from other families.

There is no universally accepted manner of making the distinction between unisensory experience and multisensory. Instead of trying to articulate such a distinction, it has been proposed to organize experiences into a hierarchy (Fulkerson, Unity, 2011). The simplest level of the hierarchy, the simplest unity, would be a mere unity in the sense that the experience is engaged by one subject. This, of course, immediately raises the problem of the unity of the subject, a longstanding philosophical problem. The point is that even to discuss the most basic kind of unified experience there is the contentious issue of what constitutes the unity of the subject.

Let us presuppose that the subject has a firm unity. What then constitutes the unity of touch? What makes touch unisensory? To get a handle on this, it is perhaps easier to look at the issue of multisensory experiences. Suppose that one hears a middle C and sees a red ball. These two experiences are easily separable. They are independent and only have an association by virtue of being in the same subject. Now, if we tighten the situation, say that the ball emits the middle C, then changing the visual orientation of the ball, moving it closer or further away, will change the auditory experience. There is interdependence. This is a multisensory experience. The information now coming to the subject is coordinated. Note that the ball itself has redness and sphericalness, and that these two properties are visual. They are not coordinated in the way that the middle C and redness are. The question is what makes the sphericalness/redness connection different from the middle C/redness connection?

The issue here is not whether touch is unisensory in the sense that the result of touch, or a feeling or sensation, is the product of one's bodily mechanism. Clearly, this is not the case. So the unity at issue in a unisensory experience is not the unity of one receptor mechanism or system. Of course, even this involves questions as to what constitutes the unity of a mechanism. Again, the unity here is not something that we can just read off the world or take for granted. For example, is my vintage Beetle car one mechanism or many? If one asserts that it is one mechanism, then it has a unity, but what constitutes that unity?

If one says that the unity is determined by function, namely, driveability, then many aspects of the car could be removed, such as doors, windows, and so on, without destroying the driveability, but it would certainly not fit the legal notion of a roadworthy car. What parts can I replace or modify without destroying the vintage nature of the car? The notion of mechanical or system unity, then, does not fall from the sky but requires a context.

So what kind of unity do unisensory experiences have? I will look at one attempt at such a characterization (Fulkerson, Unity, 2011). It has been proposed that

> unisensory experiences have a strong form of unity called "feature binding." Feature binding . . . involves the predication or assignment of distinct features to perceptual objects. The features are bound to *objects*, not each other. (p. 504)

Fulkerson does not claim that feature binding is a necessary and sufficient condition for unisensory experience, rather, that it is at least a sufficient feature. Feature binding, at first glance, is a strikingly circular process. If we ascribe redness and sphericalness to an object, the ball, and not to each other, and if the ball is nothing in experience except the conjunction of redness and sphericalness, then which one comes first: the ball or the conjunction? Fulkerson replies to this by insisting that ". . . perceptual objects are ontologically robust, objective entities" (p. 507). However, it is not clear exactly what this means except that there is a presupposition (a projection?) that underneath our experience there is a real ball, that is, the ball out there to which we fasten experienced properties: redness, sphericalness and so on. Again, these experienced elements are not attached to each other. Of course, this raises the thorny question as to how the mind reaches out and attaches perceived qualities to the real object since the real object cannot itself be perceived. It would be of interest and great support to the notion of feature binding were this binding mechanism examined and explained, but this is not yet forthcoming. In the end, this is another example of trying to individuate a sense by using an object. A single object, with features bound to it, constitutes a unisensory experience.

With regard to touch, feature binding has been less studied, but it is interesting to follow Fulkerson's discussion of it, for once again it will reveal just how much is being presupposed in the account of a unisensory view of touch. If one holds a metal sphere, one experiences a number of properties: texture, shape, roughness/smoothness, hardness and coldness. Once again it is admitted that these features may be produced along different physical mechanisms, but they are bound to the same object (p. 509). However, how can we say for sure that these tactile properties are bound more tightly or differently together than, say, the auditory and visual properties? To address this question, Fulkerson discusses ordinary experience.

> For instance, when one has a visual and auditory experience, the two experiences can be characterized individually along many dimensions: qualitative differences, different contents, different objects, etc. One can easily abstract out the auditory part from the visual part. When one has a complex haptic experience, one does not seem able to distinguish the various parts in the same way. There does not seem to be a separate kinesthetic experience independent and distinguishable from one pressure experience, both of which are different again from the thermal experience, etc. Instead, one has a unified experience with different constituent elements, as occurs in vision. (p. 509)

All of this, I would maintain, is quite true, but they are rather weak claims. However, a very strong conclusion is then drawn: ". . . a haptic experience does not involve a purely cutaneous experience that becomes associated with (or altered by) a kinesthetic experience" (p. 509). The premises are assertions involving "seems like," while the conclusion is presented as fact. All the support for such a unisensory account is essentially phenomenological: experiential.

The issue here is not that this account is necessarily wrong, but it is not one that provides a lot of philosophical comfort. This kind of phenomenological analysis is always quite risky when it comes to asserting unities in experience. In other words, experiential unity might indeed be a product of memory loss and processing speed. This was indeed recognized centuries ago by Immanuel Kant, who critiqued the notion of the *a priori* view of mathematics, which held that a simple arithmetic proposition, like 7+5=12, was analytic (Kant, 1996, p. A6-7/B10-11). Essentially, Kant argued that such a proposition was in fact synthetic. In other words, the "7+5" was not an inseparable unity with the "=12." It may appear, within experience, to be a unity, but this is simply the result of the continuous use of such simple propositions and the forgetting of their learning. Throughout the history of philosophy, many propositions have been asserted as *a priori* when in fact they have been later seen to be products of experience and are pieced together, not original unities.

With this in mind, the experience of touch as unisensory may indeed be well-entrenched in the sense that it appears as a unity, but how secure is such an appearance? It could be very similar to the unity of the mind, the self, which Hume, centuries ago, argued was illusory (Hume, 1978, pp. 251–262). The mind, Hume argued, was not a true unity, but in fact a kind of amalgamation of memories and impressions, none of which was anything more than accidentally linked. This view of the interconnected self was not something that could be absolutely asserted, but it could not be denied either. That is, the unity of the self was not something that we could prove and disprove; instead, it was a belief that one could not reject in any real sense of the term "rejection." In the end, touch as a unisensory experience could very well be similar to the experience we have of the external world: we cannot prove (or disprove) that there is a world out there, but we cannot help but believe it.

Conclusion

On the basis of the difficulties of individuating the senses and the articulation of the necessary and sufficient conditions for a sense, it appears that touch-scepticism is the position to take. From the opening discussion concerning Descartes, there is a fundamental element of confusion concerning our sensations. That is not to say that we are confused that we are sensing something; clearly, we can distinguish between thinking about a mathematical concept and feeling cold, but any attempt to, so to speak, "crawl inside" the feeling of cold and articulate all of its detail is just phenomenologically riddled with conventions. In the final analysis, touch and the other senses may seem on the surface to be so radically different, but once we try to say exactly how they are so, the difficulties begin to emerge. Touch and the senses are not as easily separable as previously thought. If so, it might well be advised that examinations of the senses, like touch, could play larger roles in general epistemological investigations.

References

Anscombe, E., and Geach, P. (Eds.). (1971). *Descartes' philosophical writings.* E. Anscombe, and P. Geach, Trans. New York: Pearson.

Aristotle. (1984). *The complete works of Aristotle: The revised Oxford translation* (Vol. I). J. Barnes (Ed.). New Jersey: Princeton University Press.

Bremier, J. (2011). Aristotle on touch. *Forum Philosophicum, 16*(1), 73–87.

Cottingham, J. (1978). Descartes on thought. *The Philosophical Quarterly, 28*(112), 208–214.

Descartes, R. (2000). *Philosophical essays and correspondence.* R. Ariew (Ed.). Indianapolis: Hackett Publishing Company.

Diaconu, M. (2006). Reflections on an aesthetics of touch, smell and taste. *Contemporary Aesthetics,* 1–14.

Fulkerson, M. (2011). The unity of Haptic touch. *Philosophical Psychology, 24*(4), 493–516.

Fulkerson, M. (2012). Touch Without Touching. *Philosopher's Imprint, 12*(5), 1–15.

Gaukroger, S. (1995). *Descartes: An intellectual biography.* Oxford: Oxford University Press.

Grice, H. (1966). Some remarks about the senses. In R. J. Butler (Ed.), *Analytical philosophy: First Series* (pp. 133–153). Oxford: Blackwell.

Hazen, A. (1990). The mathematical philosophy of contact. *Philosophy, 65*(252), 205–211.

Hume, D. (1978). *A treatise of human nature.* P. Nidditch (Ed.). Oxford: Oxford University Press.

Kant, I. (1996). *Critique of pure reason.* W. Pluhar and P. Kitcher, Trans. Indianapolis: Hackett Publishing.

Nudds, M. (2003). The significance of the senses. *Proceedings of the Aristotelian Society, 104,* 31–51.

O'Mathuna, D., Pryjmachuk, S., Spencer, W., Stanwick, M. and Matthiesen, S. (2002). A critical evaluation of the theory and practice of therapeutic touch. *Nursing Philosophy* (3), 163–176.

Perricone, C. (2006). The aspiration to the condition of touch. *Philosophy and Literature,* 30, 229–237.

Plato. (1997). Republic. In Plato, J. M. Cooper and D. Hutchinson (Eds.), *Plato: Complete works.* G. M. Grube and C. D. Reeve, Trans. (p. 1808). Indianapolis: Hackett Publishing Company.

Ratcliffe, M. (2012). What is touch? *Australasian Journal of Philosophy, 90*(3), 413–432.

Sorabji, R. (1971). Aristotle on differentiating the five senses. *Philosophical Review, 80*(1), 55–79.

Wittgenstein, L. (1958). *Philosophical investigations* (2nd ed.). G. Anscombe, Trans. Oxford: Basic Blackwell Ltd.

Chapter III

Contributions of Sensory Anthropology and Durand's Anthropology to the Symbolic Study of Touch and the Understanding of Boundaries in Psychotherapy

Christian R. Bellehumeur and Jane Chambers

At all times and in all cultures, areas of different senses of touch (i.e., thermal sensitivity, kinaesthetic, pain) have carried various meanings and have inspired many figurative expressions which fertilize the imagination in different disciplines of human thought (Coulmas, 2012). In the context of psychotherapeutics, beyond many frequently encountered expressions which are derived from touch (e.g., "It touches me"), various methods of symbolic language (i.e., metaphor, parable, etc. (Monbourquette, 1984)) involve the sense of touch to explain the complex concept of boundaries. The effectiveness of these verbal expressions depends largely on the socio-cultural context (Ricoeur, 1976) and the symbolic abilities of people involved (Durand, 2005)

We hope to demonstrate the contributions of two anthropological approaches to the study of symbolic touch as it relates to the understanding of professional boundaries in psychotherapy. This chapter has two objectives and is divided into three main sections. First, we highlight the contribution of sensory anthropology (Classen, 2005; Howes and Classen, 2013; Le Breton, 2006), which emphasizes the importance of considering the influence of culture in the study of touch (Classens, 2005, 2012; Coulmas, 2012). Second, we explore Schopenhauer's famous parable of the porcupines, which deals with the sense of touch and boundaries and offers various symbolic meanings that will help further define what is meant by healthy and unhealthy boundaries (Whitfield, 1993; Knapp and VandeCreek, 2012). We then apply the symbolic significance of touch within the complex notion of boundaries in psychotherapy to a second anthropological approach: the Theory of Anthropological Structures of the Imaginary (Durand, 1960 [1992]).[1] This approach sees the human being in terms of *Homo symbolicus* and combines the cultural influences of human experience, via reflexology, with the biological roots of the experience of the senses such as the tactile derivatives. We demonstrate how this theory, which offers three main structures (heroic, mystic, and systemic), provides a useful conceptual framework to account for existing differences between unhealthy and healthy boundaries.

The Contributions of Sensory Anthropology

The study of symbolic touch has recently evolved within the expanding field of sensory studies, notably with sensory anthropology. This interdisciplinary approach calls upon various others (i.e., history, sociology, psychology, and medicine), and involves a reciprocal relationship between the senses and culture: a cultural approach to the study of the senses and a sensory approach to the study of culture (Howes and Classen, 2013). As Classen explains:

> When we examine the meanings associated with various sensory faculties and sensations in different cultures we find a cornucopia of potent sensory symbolism. Sight may be linked to reason or to witchcraft, taste may be used as a metaphor for aesthetic discrimination or for sexual experience, an odour may signify sanctity or sin, political power or social exclusion. Together, these sensory meanings and values form the sensory model espoused by a society, according to which the members of that society "make sense" of the world, or translate sensory perceptions and concepts into a particular "worldview." There will likely be challenges to this model from within the society, persons and groups who differ on certain sensory values, yet this model will provide the basic perceptual paradigm to be followed or resisted. (1997, p. 402)

To firmly establish itself as an alternative approach to the study of culture, Classen (1997) indicates that the anthropology of the senses had to overcome three postulates which were deeply anchored in Western science. The first assumption is that the senses are "windows to the world," innate and consequently prior to culture. As Howes (2013) states, "The senses are not simply passive receptors. They are interactive, both with the world and each other. Perception is not solely a mental or physiological phenomenon" (p. 16). Sensory Anthropology argues rather that society contextually regulates all aspects of the individual's environment, including symbolic meanings. Perception is developed and acquired through cultural and personal education (Classen, 1997).

The second assumption can be overcome through the development of an anthropology of the senses that focuses on the study of senses long neglected in Western societies, which emphasize sight (and secondly, hearing) at the expense of touch, smell and taste (Classen, 1997). "The anthropology of the senses, however, argues that we must try to understand the value of each sense within the context of the culture under study and not within the context of the sensory model of the anthropologist's own culture. This means attending to the meanings encoded in all of the senses" (Classen, 1997, pp. 402–403). As Howes (2013) points out, "Each culture elaborates its own ways of understanding and using the senses. No one sensory model will fit all" (p. 16). That is why sensory anthropologists use an ethnography based on experience rather than solely on writings, while questioning the impact of a literary-minded approach which may suppress or transform the dynamic multi-sensory dimensions of culture (Classen, 1997).

The third assumption to overcome is owed, ironically, to academics who have questioned the visual hegemony within the study of culture. "Marshall McLuhan (1962) and Walter Ong (1967), notably, argued that the sensory model of a society is determined by its technologies of communication. According to this theory, literate . . . societies emphasize sight due to the visual nature of writing, while non-literate societies emphasize hearing due to the auditory nature of speech" (Classen, 1997, p. 403). However, this new methodological approach to culture, says Classen (1997), encapsulated cultures within either visual or auditory sensory models. Whereas "the sensory anthropologist attends to the role of odours, tastes and tactilities—as to the role of sights and sounds—not as evidence of evolutionary status, nor as picturesque detail such as may be found in a travel guide, but as essential clues to the ways by which a society fashions and embodies a meaningful world" (Classen, 1997, p. 405).

To study senses and their meanings in the context of sensory anthropology, Howes and Classen (1991) have elaborated 10 main methodological categories (here described using the sense of touch): (1) verbal expressions (e.g., "this touches me"); (2) artifacts and aesthetics (e.g., fashions showing various parts of the body, waxing); (3) bodily decorations (e.g., tattoos); (4) children's education (e.g., North American societies tend to restrain touching children in schools); (5) alternative sensory modalities; (6) communications media; (7) natural and artificial environments such as the virtual world (e.g., internet either fosters communication, or replaces real human interaction and touch); (8) rituals (e.g., Catholic liturgies are mainly visual and auditory experiences); (9) mythology; and (10) cosmology. Illustrating the latter, the Judeo-Christian paradigm emphasizes the Greek mind-body dualism: the underworld's (earth) carnal passions versus the upper-world's (heaven) moral salvation. The same can be said for the scientific paradigm (mainly Cartesian thinking), which now influences our ethics using the verbal schema of separation (e.g., between professional and personal).

To illustrate the influence of culture on our understanding of the experience of touch, let us look at a famous story about touch and boundaries from the nineteenth century.

Schopenhauer's Parable of the Porcupines

> A company of porcupines crowded themselves very close together one cold winter's day so as to profit by one another's warmth and so save themselves from being frozen to death. But soon they felt one another's quills, which induced them to separate again. And now, when the need for warmth brought them nearer together again, the second evil arose once more. So that they were driven backwards and forwards from one trouble to the other, until they had discovered a mean distance at which they could most tolerably exist. (Schopenhauer, 1851, pp. 651–652)

This parable is interesting because it is based on the sense of touch and provides various meanings. The first meaning is that of the pain and heat felt by the porcupines' skin, by which the philosopher clearly communicates the inherent difficulties of human contact, of human touch.

To explore these complexities and their cultural influences, there are three characteristics to consider in the phenomenology of touch: they are gestural, impactful, and reciprocal (according to Leder and Krucoff, 2008).

Gestural. "To see or hear something, it can be sufficient to wait motionless, . . . by contrast, touch, when used to explore the world, relies upon an active movement, a gesture" (Leder and Krucoff, 2008, pp. 323–324). Furthermore, as sensory anthropology points out, the influence of one sense cannot be isolated from the influence of other senses. According to Saint-Arnaud (2002), the richest pleasures linked to the ability to heal come from kind gestures and words (cf. hearing) from loved ones. The pleasure and the security which can be offered by touch, both verbal and tactile, remain a powerful healthy and healing factor (Saint-Arnaud, 2002). This is thanks to the sound effect of words which produce first a physical touch: on the ear, but also soothing muscles they meet through the skin. We all know that some words really do make us shiver with pleasure (Saint-Arnaud, 2002).

Impactful. Also from Leder and Krucoff (2008), touch depends upon physical impact, contact between bodies, blended with the experience of force. "This explains the great cultural significance attached to the loving or consoling touch, the sexualized touch, or the invasive and hostile touch, all of which are carefully regulated by social codes" (Leder and Krucoff, 2008, p. 324). This cultural influence extends to cosmology in sensory anthropology. Touch is one of only two of our five senses (along with taste) which need immediate contact with the object; sight and hearing have thus been given a curious dignity in the West, where touch is deemed impure because the contact always involves the physical or material world (Coulmas, 2012).

Reciprocal. Lastly, to touch someone implies being touched, which refers to reciprocity. Leder and Krucoff (2008) argue that "[t]ouch's intimacy of contact can make us aware of the reciprocity of inner and outer, as well as that of body and the world, self and other. Rather than perceiving the world across a distance, touch is almost literally defined by the surrender of boundaries and the adoption of relational intimacy" (Leder and Krucoff, 2008, p. 324). According to Coulmas (2012), for the child, the tactile act provides an "external" perception, as well as an "internal" perception. The inherent reflexivity in this sense serves as a model to all other sensory reflexivities (e.g., seeing oneself in a mirror and observing oneself doing that). These, in turn, generate the reflexivity of consciousness. Thus, the first discovery, the first communication is done by touch, the primary sense. This also implies later that the first forbidden and early struggles are related to this sense of touch. During childhood, explorations will inevitably come with the warning. "do not touch," because touching can be associated with hurting oneself, or other people or things. These warnings are conditioned by culture, just like the proper psychological distance relies on cultural codes (Hall, 1990).

Schopenhauer's parable has a second level of meaning, which refers to the notion of social and interpersonal boundaries: the challenge of maintaining proper balance between painful entanglement and loveless isolation.

> Thus the need for society, which springs from the emptiness and monotony of men's lives, drives them together; but their many unpleasant and repulsive qualities and insufferable drawbacks once more drive them apart. The mean distance which they finally discover, and which enables them to endure being together, is politeness and good manners. Whoever does not keep to this, is told in England to 'keep his distance.' By virtue thereof, it is true that the need for mutual warmth will be only imperfectly satisfied, but, on the other hand, the prick of the quills will not be felt. Yet whoever has a great deal of internal warmth of his own will prefer to keep away from society in order to avoid giving or receiving trouble and annoyance. (Schopenhauer, 1851, pp. 651–652)

Hence, this parable is referenced in Freud's own writing (1921), as it resonates with his determinism and his pessimistic view of human nature.

> Let us keep before our eyes the nature of the emotional relations which hold between men in general. According to Schopenhauer's famous simile of the freezing porcupines no one can tolerate a too intimate approach to his neighbor. The evidence of psycho-analysis shows that almost every intimate emotional relation between two people which lasts for some time—marriage, friendship, the relations between parents and children—contains a sediment of feelings of aversion and hostility, which only escapes perception as a result of repression. (p. 101)

In our view, Schopenhauer's and Freud's views on human boundaries are rather pessimistic and belong to the authors' subjective and ethnocentric mindsets, along with their narrow use of the characteristics of touch. Problems avoided by three key aspects of sensory anthropology are first, while studying any sense, one needs to take into consideration the cultural dimension, which largely conditions any sensory experience; second, sensory anthropology cautions against any cultural bias; and third, sensory anthropology clearly states that the study of one particular sense cannot be done in isolation from other senses. In the following section, we will see that these three main points are all considered in another theoretical approach, Durand's anthropological structures of the imaginary.

The Contributions of Durand's Anthropology of the Imaginary

Since his seminal book (1960), philosopher and anthropologist Gilbert Durand has shown, through works in aesthetics, music and literature, philosophy, religious mythology, esotericism, schools of humanities, and hermeneutics of applied science, how human activity unfolds in various regimes (systems) or fundamental structures of the imaginary. Durand's (1960 [1992]) Theory of Anthropological Structures of the Imaginary considers human beings as *Homo symbolicus*, and their symbolic world as "the entire human universe" (Durand, 1979). There, he said, lies the origins of human culture, namely in a wealth of images and symbols that continue to shape our ways of thinking, living, and dreaming (Xiberras, 2002). These images are organized in basic dynamic systems that give them a particular meaning: "The images are raw psychic forces [that] are stronger than ideas, stronger than real experiences" (Bachelard, 1948, p. 20). These reservoirs of images somehow represent common archetypal patterns that unfold in an "anthropological trajectory," according to which there is a bilateral influence between physiology and society (Durand, 1960 [1992]). Therefore, according to Wunenburger (2013), imagination is no longer just a rational position but is part of the whole constitution of human living. In his theoretical approach, Durand (1960 [1992]) combines not only the cultural influences of human experience, but also takes into account the biological roots (via reflexology) of the experience of the senses, such as the tactile aspect.

According to Durand (1960 [1992]), the deployment of these archetypal patterns generates two large "regimes" or systemic ways of organizing mental images and visual narratives, referred to as diurnal and nocturnal. These schemas are based on opposing visions of the world: one structure of the imaginary is identified as "heroic and purist," and the other as "intimate and enmeshed" (or mystical). Only a third category of structures, called "synthetic" (which Durand later renamed "systemic"), can enable the two opposing regimes of the imagination to achieve a healthy harmony while remaining in each other's presence, instead of excluding one another.

The following sections briefly describe each system, including how it shapes the assembling of images according to its own logic. Table 1 is a quick overview of Durand's classification of images based on dominant reflexes and senses.

Table 1: Durand's Classification of Images

Orders or polarities	Diurnal	Nocturnal	
Structures	Heroic (or schizomorphic)	Synthetic (or dramatic or systemic)	Mystical (or intimist)
Verbal schemata	Distinguish	Link	Confound or mix
Dominant Reflexes	POSTURAL dominant with its manual derivatives and the support of sensations at a distance (sight, hearing)	COPULATIVE dominant with its rhythmical motor derivatives and its sensorial support (kinetic, rhythmo-musical, etc.)	DIGESTIVE dominant with its coenaesthetic and thermic supports and its tactile, olfactory and gustatory derivatives.

The Mythical World of the Mystical Type

The "intimate/mystical" structural division is linked to the nocturnal regime of the image (Durand, 1960 [1992]). Durand here places the sense of touch, which he calls tactile derivatives. This regime is about finding peace, without turbulence; everything is fusion, friendly, gentle, harmonious and peaceful. Life is warm while internalized, like good food that we assimilate to sustain ourselves. Descending or swallowing gestures characterize the action of ingesting, the basic reflex of this regime (Xiberras, 2002). The human becomes a single entity with the food he or she eats, as food is lost in the depths of his or her body. It is the world of intimacy, privacy, and absorption, just like the skin absorbs substances. In this intimate ambiance, people (i.e., in psychotherapy) explore their feelings (e.g., taste, touch), and spouses seek harmony, sweetness and affection from a loving embrace.

This regime carries a sort of relational energy that brings things and people in harmony in an atmosphere of intimacy, closeness, and solidarity. This binding energy echoes the principles of analogy and similarity with the predominant verbal schema "to confound or to mix." In the mystical polarity, it is important that each person feels good and tries to create an environment of trust and safety for one another. In the helping relationships, if the secret garden of the one is revealed or known, the other respects and treasures it. Indeed, discretion and confidentiality are valued when what is shared must remain "between us." This is most important in the context of boundaries: just like food is necessary for our survival, the absence of the sense of touch causes the death of living beings (Coulmas, 2012).

The Mythical World of the Heroic Type

At the opposite end, the "heroic" or "purist" structural division (or regime) of daytime image is characterized by the principles of separation, identity, distinction, exclusion and contradiction (Xiberras, 2002). The basic reflex of this regime is the act of getting up or standing, and the verbal schema is one of "distinguishing (i.e., differentiating)" (Durand, 1960 [1992]). In this regime, symbols refer to making a stand: separating what is good from what is bad or evil, or what makes us rise from what makes us fall, to battle and victory (Xiberras, 2002).

On a relational level, linked to boundaries and the helping relationship, one can conceptualize the heroic world in terms of relational energy that recalls the importance of individuality, autonomy, and identity (Laprée, 2013). As with sight and hearing, it is important to keep a healthy distance with what is different from us. It is because one can clearly distinguish his or her limits and strengths that he or she can establish clear boundaries with others. This reflects the tendency to protect ourselves, to keep our identity consistent, as the hairs of the skin that stand up indicate that the pores of the skin are closed to protect against unwanted foreign bodies.

The Mythical World of the Synthetic Type

In an individual or in a community, if an imaginary system (regime) takes up too much space at the other's expense, the latter regime becomes lethargic, an unhealthy state grows and causes discomfort (i.e., personality disorders, deep-rooted cultural conflicts). Overcoming these problems, the "synthetic" structural system is characterized by the principle of causality and the balanced coexistence of the two previously discussed, opposing structural poles (Durand, 1960 [1992]). In the synthetic regime, the verbal schema is "to link" or "to put together" the opposites in a dramatic, steady rhythmic or cyclical way (Xiberras, 2002).

From a biopsychological point of view, the synthetic structural division requires diving into ourselves and becoming aware of our mortality in order to make good use of time. It is rooted in the gestures of the act of mating, considered as the basic reflex of this regime, where male and female unite and become a whole, while remaining separate and opposite entities. From a cultural point of view, the synthetic mythical world includes different systemic structures that are found in various cultural traditions and myths, which implement the *coincidentia oppositorum* (or coexistence of opposites) (Xiberras, 2002). These structures unfold by pursuing the ideas of progress, growth, health, mental balance, or by recycling essential and fundamental elements of the past into the present in a continuous motion, such as by the process of storytelling (Laprée, 2000).

For example, the rhythm patterns of different musical styles allow for the control of time because of their stabilizing function, which either stimulates or soothes us. The aging process of the skin reflects the process of historicity and maturity, which belongs to the synthetic universe (Durand, 1960 [1992]). Thus, the synthetic embraces a paradox: the strength of age (cf. wisdom) versus its weakness (the skin's loss of elasticity and beauty). Expressions such as "being born again" also belong to the synthetic regime. So do symbols such as fire because it can both burn and warm. For Coulmas (2012), opposites coexist through the sense of touch and its organ, the skin, which is the only sense organ to get dirty, either concretely (i.e., direct contact with repulsive matter), or figuratively (i.e., immoral acts) (Coulmas, 2012). The "clean" and "dirty" problem is further complicated by the religious origin of the "pure" and "impure" quality of the skin, distinguishing the sacred from the profane (Coulmas, 2012). This notion remains in the background of our reasoning and is critical for understanding underlying motives of hygiene.

In short, because of the complexity of boundaries and based on concepts from sensory anthropology, we cannot simply rely on Schopenhauer's parable to define boundaries, particularly in the context of psychotherapy (Barnett, 2014).

Defining Boundaries in Human Life and in the Context of Psychotherapy

Boundaries can be defined as limits: they distinguish between self and others; separate what belongs to oneself from what does not; identify what makes oneself unique (Whitfield, 1993). For example, the skin tracks the physical boundary of our body but also our psychological space. Boundary is not synonymous with isolation, or lack of need or desire to love and be loved. A boundary is as far as one can comfortably go in a relationship (Whitfield, 1993). Our freedom ends where the other's begins. We can enable some people to approach us, physically and emotionally, and keep others away. So when we manage our own boundaries, contact with others becomes an expression of self, integrity and freedom (Whitfield, 1993). The challenge with boundaries is to become aware of the impact of others on ourselves, and of our own impact on others. Thus, while touch is intrinsically linked to contact, making contact with others means doing so tactfully and with respect for our own and others' boundaries.

Whitfield (1993), a medical doctor, describes boundaries using the role of human cells. Each cell has a semi-permeable membrane and lives among other cells. When the cell works well, it keeps toxins (dangers) on the outside, allowing the nutrients inside and eliminating the surplus after digesting. We also have the intelligence and resources to function as a cell—that is, to know which relationships are nourishing and which are too invasive and toxic. Being unaware of our own cell, its contents and its operation, puts us at risk of being too flexible and adaptable, thus allowing others to abuse us.

This biological metaphor introduces the idea that boundaries can be healthy or unhealthy. Healthy boundaries mean protection, clarity, firmness, maintenance, appropriateness, receptivity, mutual reciprocity, flexibility, awareness in the present, relevance of inner life, adaptability, and self-control (Whitfield, 1993; Knapp and VandeCreek, 2012). Unhealthy boundaries refer to distance, isolation, rigidity, manipulation and control, alienation, absence or nonexistence, enmeshment, fusion, determination by others, and triangulation (Whitfield, 1993; Knapp and VandeCreek, 2012).

In any interpersonal relationship, and particularly in the context of psychotherapy, the acknowledgement and respect of healthy boundaries are fundamental. "Each of the mental health professions makes very clear in their codes of ethics the importance of establishing and maintaining clear boundaries between clinicians and those they serve (e.g., clients, supervisees, and students) to prevent actions and behaviours that may lead to exploitation of, or harm to, these individuals" (Barnett, 2013, p. 170). Three of these boundaries in the context of psychotherapy (along with location, time and gifts) are directly or implicitly discussed in this chapter: interpersonal space, self-disclosure, and touch (Barnett, 2013). Nevertheless, although the understanding of what are healthy and unhealthy boundaries may be clear for some people, at least at a conceptual level, there may be various (cultural, personal, etc.) factors which make them not so easy to fully understand and honour in real-life settings. At times, there may be grey zones that make cut and dried distinctions about healthy and unhealthy boundaries somehow arbitrary. Therefore, we suggest that the complexity surrounding boundaries can be further explained by Durand's framework.

Application of Durand's Theory to the Understanding of Boundaries

Gilbert Durand's *Theory of Anthropological Structures of the Imaginary* proposes an encompassing conceptual framework that attempts to link the subjective view of the phenomenology of touch and the more objective view of cultural contributions from sensory anthropology. To elaborate our view, we provide five key points below.

First, to demonstrate the relevance of Durand's theory to the understanding of touch and to the concept of boundaries in interpersonal and psychotherapeutic relationships, we must clarify the relational needs involved. There are two main human relational needs: maintaining one's individuality (identity or autonomy), and seeking closeness and intimacy (Meier and Boivin, 2011). It is understood that even if professional (psychotherapeutic) relationships do not imply the same nature of intimacy as that of a friendship, or romantic or conjugal relationship, all types of interpersonal relationships do imply a certain level of intimacy in order to foster trust and remain alive on both sides.

Since these two basic relational needs are essential, a delicate question is often raised: how to simultaneously develop identity while deepening intimacy? Different authors suggest that "the goal of life is to become a mature adult who has achieved a sense of differentiation of self . . . , who has learned how to balance the two life forces of individuation and togetherness" (Rovers, 2006, p. 239). However, depending upon our particular level of differentiation along the individuation-togetherness continuum (Bowen, 1978), one may wonder how to reconcile these two basic relational needs (identity versus intimacy). As Covey (2011) stated, in any binary model (e.g., I versus You; identity versus intimacy), problems arising from tensions between two poles need to be resolved by imagining a third path. We argue that Durand's (1960 [1992]) theory answers this question by looking at problem-solving through the third path of allowing the coexistence of opposites within the synthetic structure (Wunenburger, 2013), where themes of maturity, historicity, growth, and progress belong.

Second, the question raised above refers to the challenge of regulation (St-Arnaud, 1989). Individuals and couples go into counselling mainly because they do not know how to manage their conflicts. According to Gottman and Levenson (2002), the way in which couples manage their conflicts will determine whether they will remain together.

Third, in Durand's (1960 [1992]) theory, the two life forces of individuation and togetherness are included in two opposite regimes: the heroic structure relating to the need of individuation or identity, and the mystical structure involving the need for intimacy or togetherness. Whereas partners' different relational needs vary from one pole to the other, both of these psychological concepts may cohabit and are inherent to the maintenance of healthy boundaries.

Fourth, even if the sense of touch (tactile derivatives) is classified in the mystical/mythical regime (see Table I), the experience of boundaries may involve other senses in various ways, based on principles of sensory anthropology (Classen, 1997).

Fifth, we argue that resolving conflicts often implies dealing with boundaries. For example, while one romantic partner might demand greater intimacy in the relationship, the other feels that greater intimacy suffocates his or her personal identity. According to their level of self-differentiation, partners may suffer unmet relational needs for different reasons.

To illustrate our view, one can see in Table 2 how Durand's theory provides a useful framework to account for differences between unhealthy (rigid or in fusion) and healthy boundaries. All characteristics of healthy boundaries require both positive heroic and mystical polarities (while their negative forms correspond to unhealthy boundaries). Therefore, healthy boundaries echo the third mythical universe, the synthetic (or systemic) structure, which organizes images and symbols to maintain a clear identity while encouraging openness to intimacy, thus allowing heroic and mystical structures to coexist.

Table 2: Applications of Durand's Categories of the Imaginary on Boundaries' Characteristics

Heroic Polarity (healthy or positive)	Mystical Polarity (healthy or positive)	Heroic Polarity (unhealthy or negative)	Mystical Polarity (unhealthy or negative)
Protective	Receptive	Distant	Absent or almost
Clear	Mutually reciprocal	Isolated	Tangled or diffuse
Firm	Flexible	Rigid	Determined by others
Robust (insuring maintenance, coherence)	Aware of now / Issuing from an inner life	Manipulating others or controlling others	Fusion
Appropriate	Adapted	Alienated or insane	In triangulation

Conclusion

We sought to demonstrate the contributions of two anthropological approaches to the study of symbolic touch and the understanding of professional boundaries in psychotherapy. Among the various points raised by sensory anthropology, three main ones are taken into account in Durand's theory of imaginary: (a) the influence of culture is of great importance to the study of the experience of touch; (b) cautiously avoid any ethnocentric point of view when studying senses; (c) while studying any particular sense, all senses need to be taken into consideration, as they influence each other.

Beyond its theoretical contribution, Durand's theory may be useful from an ethical perspective. In the context of psychotherapy, as Barnett (2013) pointed out, boundaries can be avoided, crossed, or violated. Historically, "it was believed that all boundary violations, and even all boundary crossings, placed the psychotherapist at risk of . . . harm(ing) . . . the client. Accordingly, it was strongly recommended that psychotherapists take a very conservative view of boundaries and be very cautious about ever crossing them" (Barnett, 2013, p. 174). This position clearly corresponds to a polarized view (i.e., heroic structure) of the imaginary: A psychotherapist keeps an extreme (i.e., distinguishing) distance from his or her clients, out of fear of causing harm, without looking at ways they may relate to each other. However, according to Lazarus and Zur (2002), there is no clear evidence suggesting that crossing one boundary will ultimately lead to violating others (e.g., a psychotherapist ending up having a sexually intimate relationship with one of his or her clients).

> There are numerous occasions where psychotherapists engage in boundary crossings that are clinically appropriate and that do not lead to subsequent violations . . . (such as) the therapeutic efficacy of disclosure of self . . .. Not only is a rigid and avoidant approach to boundary crossings uncalled for, but in many situations . . . would in fact be harmful to the client and to the psychotherapy relationship . . . such as a refusal to touch a grieving client . . .. (Barnett, 2013, pp. 174–175)

Psychotherapists need to be flexible in approaching boundaries, as the context and nature of the client/therapist relationship may vary greatly. Psychotherapists may knowingly choose to live certain paradoxes where opposites coexist. Such boundary-crossing clinical situations correspond to Durand's systemic structure of the imaginary where firmness and flexibility coexist. Blindly applying a pre-established code of conduct would be unethical and, indeed, sterile for the development of the individual.

In regards to Schopenhauer's parable of the porcupines, beyond the two meanings previously discussed, namely, the sense of touch, and the notion of boundaries, there may be a third meaning underlying this parable. There is something not immediately visible yet present in any human relationship, a sort of personal and social energy (de Peretti, 1999) which is vital in any interpersonal and therapeutic relationship. Once people come together, fields of perceptible energy are created which have an impact on people's (e)motions. Any presence necessarily creates opportunities for friction, multiple rapports, difficulties, and detours for everyone involved. Any relationship is energetic and implies a dynamic regulation which was discussed in this chapter, referring to St-Arnaud (1989) and Durand's theory of the imaginary.

Finally, if we may rarely physically touch our clients in a therapeutic session, we can, at the very least, discuss this basic need. Both psychotherapist and client can certainly be emotionally and spiritually touched, as a world of images and energies circulate in any given relationship. This cultural world of images is filled with emotions which live within us, as illustrated by Gilbert Durand's anthropological trajectory of *Homo symbolicus*. It organizes our human relations and helps us regulate our boundaries.

Note

1. In his work, Durand's (1960 [1992]) mapping has its roots in sensorimotor schemas, namely three dominant reflexes (i.e., dominant posture, descending or digestive gesture, and rhythmic gesture) (Xiberras 2002), which fall within the anthropological and are linked to the neurobiological and to the psycho-spiritual. As presented in Table I, the three main mythical structures have corresponding verbal schemata and dominant reflexes. It is interesting to note that Durand puts the two senses, sight and hearing, seen as dominant ones in the West, in the heroic regime. Sensory anthropology has emerged as a way to compensate for visual supremacy, which Durand calls "heroic." While both vision and sound imply the notion of distance, touch and taste are associated with the notion of rapprochement or closeness and the corresponding verbal schemata confound or mix (which can lead to fusion).

References

Bachelard, G. (1948). *La Terre et les rêveries du repos*. Paris: José Corti.

Barnett, J. E. (2014). Sexual feelings and behaviors in the psychotherapy relationship: An ethics perspective. *Journal of Clinical Psychology: In Session, 70*(2), 170–181.

Bowen, M. (1978). *Family therapy in clinical practice*. New York: Aronson.

Classen, C. (1997). Foundations for an anthropology of the senses. *International Social Science Journal, 49*(153), 401–412.

Classen, C. (2005). *The book of touch*. Oxford: Berg.

Classen, C. (2012). *The deepest sense: A cultural history of touch*. Illinois: University of Illinois Press.

Coulmas, C. (2012). *Le toucher (Métaphores des cinq sens dans l'imaginaire occidental)*. Les Éditions La Métamorphoses.

Covey, S. R. (2011). *The 3rd alternative*. Franklin Covey Co.

de Peretti. A. (1999). Énergétique personnelle et sociale, coll. Cognition et formation. Paris: L'Harmattan.

Durand, G. (1960 [1992]). *Les Structures anthropologiques de l'imaginaire : Introduction à l'archétypologie générale*. Paris: Dunod.

Durand, G. (1979). *Figures mythiques et Visage de l'œuvre: De la mythocritique à la mythanalyse*. Paris: Berg International.

Durand, Y. (2005). *Une technique d'étude de l'imaginaire: L'AT.9*. Paris: L'Harmattan.

Freud, S. (1921). Group psychology and the analysis of the ego. Chapter VI – Further problems and lines of work. The standard edition of the complete psychological works of Sigmund Freud (Translation: James Strachey), Vol. XVIII, p. 101.

Gottman, J. M. and R. W. Levenson. (2002). A two-factor model for predicting when a couple will divorce: Exploratory analyses using 14-year longitudinal data. *Family Process*, *41*(1), pp. 83–96.

Hall, E. T. (1990). *The hidden dimension*. Reissue edition. Anchor.

Howes, D. et C. Classen (1991). Sounding sensory profiles. In D. Howes (Ed.), *The varieties of sensory experience: A sourcebook in the anthropology of the senses*. Toronto: University of Toronto Press.

Howes, D. (2013). *The expanding field of sensory studies*. Sensory Studies. Retrieved from www.sensorystudies.org.

Howes, D. and Classen, C. (2013). *Ways of sensing: Understanding the senses in society*. London: Routledge.

Knapp, S. J. and L. D. VandeCreek. (2012). *Practical ethics for psychologists: A positive approach* (2nd ed.). Washington, DC: APA Books.

Laprée, R. (2000). *La psychagogie des valeurs. Symbolique et imaginaire en éducation*, Montréal, Éditions Logiques.

Laprée, R. (2013). La dynamique de groupe et les structures anthropologiques de l'imaginaire. In R. Laprée and C. R. Bellehumeur (Eds.), *L'imaginaire durandien – Enracinements et envols en Terre d'Amérique* (pp. 3–17). Québec: Presses de l'Université Laval.

Lazarus, A. A. and O. Zur. (2002). *Dual relationships in psychotherapy*. New York: Springer.

Le Breton, D. (2006). *La Saveur du monde: Une anthropologie des sens*. Éditeur Métaillé.

Leder, D. and M. W. Krucoff. (2008). The touch that heals: The use and meanings of touch in the clinical encounter. *The Journal of Alternative and Complementary Medicine*, 14(3), 321–327.

Meier, A. and M. Boivin. (2011). *Counselling and therapy techniques: Theory and practice*. Washington, DC: Sage.

Monbourquette, J. (1984). *Allégories thérapeutiques, ou Histoires pour instruire et guérir*. Ottawa, ON: Université Saint-Paul.

Ricoeur, P. (1976). *Interpretation theory: Discourse and the surplus of meaning*. Fort Worth, Texas: The Texas Christian University Press.

Rovers, M. (2006). Couple therapy with severely conflicted partners: A synthesis of Bowen and attachment theory. In A. Meier and M. Rovers (Eds.), *Through conflict to reconciliation* (pp. 239–265). Montreal, Quebec: Novalis Publishing Inc.

Saint-Arnaud, Y. (2002). *La guérison par le Plaisir*, Montréal, Québec : Novalis.

Schopenhauer, A. (1851). Parerga and Paralipomena, Vol. 2 Chapter XXXI – Similes, Parables, and Fables, #396 Translation: E. F. J. Payne. Oxford University Press, pp. 651–652.

St-Arnaud, Y. (1989). *Les petits groups. Participation et communication*. Montréal: Presses de l'Université de Montréal et Les Éditions du CIM.

Whitfield, C. L. (1993). *Boundaries and relationships. Knowing, protecting and enjoying the self.* Deerfield Beach, Florida: Health Communications, Inc.

Wunenburger, J.-J. (2013). L'anthropologie de l'imaginaire selon Gilbert Durand: Contextes, options, enjeux. In R. Laprée and C. R. Bellehumeur (Eds), *L'imaginaire durandien – Enracinements et envols en Terre d'Amérique.* (pp. 3–17). Québec: Presses de l'Université Laval.

Xiberras, M. (2002). *La pratique de l'imaginaire. Lecture de Gilbert Durand*. Laval: Les presses de l'Université Laval.

Chapter IV

Healing and the Forbidden Touch: A Reflection on Selected Scripture Stories

Karlijn Demasure

Introduction: On the Meaning of Touch

Touch is one of the five senses. The senses enable us to experience what happens outside the person. They are the means to get knowledge about the world and the Other. This chapter begins with a brief approach on the meaning of touch from a philosophical point of view and then explores two Christian biblical texts in which touch plays a crucial role. Both the text about the touch of the hemorrhaging woman in Mark 5: 25–34 and the text about the non-touching Mary Magdalene in John 20: 11–18 tell a story about Jesus being touched by women. Each woman takes the initiative, but only one is allowed to touch. Each narrative is followed by a reflection.

The aim of this conversation between the philosophical insights and the texts is to deepen our understanding of touch in the biblical stories as well as in daily life. This is an existential hermeneutical reflection, which will contribute to the construction and better understanding of the meaning of touch and will move between what has already been said, what can be added, and what can be said differently. Such a hermeneutical approach lays the foundation for interdisciplinary research and is based on Ricoeur's hermeneutics, which states that "to explain more, is to understand better" (free translation from present author) (Thomasset, 1996, p. 11; Ricoeur, 1983, 1986; Demasure, 2006; Santedi and Malou, 2006; Demasure and Muller,

2006). Each interdisciplinary research enriches one's understanding, again and again.

From Sight to Touch

In philosophy, one traditionally distinguishes five senses: vision, hearing, smell, taste, and touch. In the Western philosophical tradition, rooted in Greek philosophy, the sense of vision has always been prioritized (Wyschogrod, 1980). Plato considered the eyes to have the ability to convey the most truthful perception and to give way to the highest ethical capacity. Because vision was considered to be the highest sense, it was also closest to the divine (Kambaskovic and Wolfe, 2014). The eye and sight are bound together by light, and the sun is the author of light. If something is in the dark, you cannot see it; however, if the sun is shining on it, that thing becomes visible. The sun is the cause of vision. Plato then uses this image to talk about the soul. The soul is like the eye: "When its gaze is fixed upon an object irradiated by truth and reality, the soul gains understanding and knowledge and is in possession of intelligence" (Plato, 1945, pp. 507–508).

However, it is Aristotle who has influenced most of our thinking about the senses. He studied the senses in a very detailed way. At the very beginning of his *Metaphysics* he writes: "All men [*sic*] by nature desire to know. An indication of this is the delight we take in our senses; for even apart from their usefulness they are loved for themselves; and above all the sense of sight. . . .we prefer seeing (one might say) to everything else" (Aristotle, 2001, p. 689). Sight is the sense that, in the opinion of Aristotle, makes one know and brings to light many differences between things. It allows one to simultaneously perceive things (Jonas, 1982). Watching allows the observation of a wide visual field. In this sense, seeing is considered to be less temporal than hearing or touching. Sight is the apprehension of great distances. Therefore, in opposition to touch, sight allows a man or woman not to have to engage immediately with the object on which he or she is gazing.

Although seeing remains the most important sense, touch has a very important place as well. Aristotle defines taste and touch as tactile senses that are closest to the animal world. Taste is touching with the tongue and is thus a particular form of touching. It is not because touch and taste are closely related to the animal world that they are not important. On the contrary, humans have the finest tactile sense, and this is indicative of their superior intelligence. Touch is the very condition for being alive. While one can live without seeing or hearing, without touch neither animals nor human beings can exist (Aristotle, 2001; Derrida, 2000). Not only would humans not be able to survive because they would not feel the heat of fire, for example, but if they are not touched by the mother after birth, they would die. Both touch and taste are necessary for finding and recognizing food and thus the condition for staying alive. Touch is the sense that refers to life and death.

The purpose of the other senses is related to well-being. Touch is not intended to promote the well-being of the human person; rather, it is a precondition of both human and animal existence. Nevertheless, the awareness of pleasure and its opposite, pain, should not be ignored (Derrida, 1993). Derrida (1993) argues, "Aristotle uses the word *aphe*, which also means tact, grasp, as well as the place of contact, line of joining, blow or wound" (p. 127). Thus, the body is equipped to detect those elements that are either destructive or conducive to the well-being of the human person. Unpleasant things are infrequently hazardous and can lead to the destruction of the human person. The sense of touch includes the notion of vulnerability in the discourse.

The dominance of sight does not imply, however, an agreement among scholars (Jay, 1993). In historical thought, different streams and movements acting against this dominance of sight are indicated. Democritus, the Greek philosopher who lived around 460–370 BCE, compiled the list of five senses and, according to Bremer (2011), was probably the first to reduce all senses to touch. English (1915) notes, "It is by some kind of physical contact that all the senses operate" (p. 74). Genuine knowledge was based on perception through the senses but had to be reasoned upon. English (1915) also argues that knowledge could not be delivered directly by the senses since they were not able to "see" the ultimate matter, which were the atoms. Indeed, according to Democritus, the senses only provide us with obscure knowledge. Real knowledge was still a product of the mind, and because of that divine things are thoughts of the mind (English, 1915).

The most frontal attack on the primacy of sight has to be situated in the twentieth-century French philosophy of Nancy (2003) and Derrida (2000, 1993). Today, great importance is given to touch and consequently to the body, thus creating the impression that no longer sight but touch is given the status of paradigm (Derrida, 1993). Philosophers such as Irigaray (1993) have stressed that touch, which is linked to the skin, is the first sense to be developed in the mother's womb. Vision, according to Irigaray (1993), is privileged by men. Vision objectifies, creates distance and dominates, placing women in a passive position. Women thereby become the object of contemplation. In Irigaray's opinion, women prefer touch to vision. Touch is thus related to women, giving rise to a feminization of touch (Winter, 2001). Winter (2001) notes: "Touch is implicitly feminized in this understanding of tactility as corporeal, sexual, unreliable, and unreasonable" (p. 2).

Both sight and touch lead to knowledge, although the nature of the knowledge acquired in each instance is different. According to Vasseleu (1998), vision provides the distance necessary for objectivity and thereby makes science possible. Touch, on the other hand, furnishes knowledge through subjective immediacy, whereby qualitative difference can be observed, but also through intuitive irrefutability. Wyschogrod (1980) states clearly: ". . . that touch provides the warranty for veracity since tactile apprehension requires no intermediary; that we aspire to touch because we aspire to a truth that cannot be vouchsafed through representation . . . and thus retains its alterity" (p. 196). While sight can be misleading and falsely give the impression that something is out there, touch cannot. It definitely proves the reality out there. That is why it is the best sense to be aware of alterity. Irigaray (1993) adds that some types of touch must be recognized as objectifying the other, and the other is appropriated as an object and not recognized as another human person. Here the difference is nullified, and he or she is included in "the same" (p. 204). Irigaray (1993) refers to male sexuality where women are only considered as an envelope for men. Thus, while sight can objectify women, touch can as well.

To Touch or Not to Touch?

In the second part of this chapter, we explore the biblical text on the hemorrhaging woman (Mk 5: 25–34) and the text of Mary Magdalene at the scene of the resurrection (Jn 20: 11–18). Both texts are about touch; both are about women who want to touch; and in both cases Jesus/Christ is taken by surprise. In the former, the woman succeeds in touching Jesus and gets healed, while Mary Magdalene is not allowed to touch Christ. Beginning with the patristic times, these two scenes are often seen in opposition: the hemorrhaging woman was allowed to touch because she believed, while Mary Magdalene lacked faith. In visual arts these two scenes often get confused or intermingled. (Baert, Kusters and Sidgwick, 2012). Others, however, will recognize that Mary Magdalene lived a conversion and that she did not need to touch anymore. She accepted that Jesus has become Christ and that a different relation was needed. By this conversion, she also became a model of faith.

Touching Jesus and the Hemorrhaging Woman (Mk 5: 25–34)

For the translation of this passage, I use *The Jerusalem Bible, Reader's Edition* (1966). For the text from John, I will use the *King James Bible* (1999). The text states:

> 25 Now there was a woman who had suffered from a hemorrhage *for twelve years;*
> *26 after long and painful treatment under various doctors, she had spent all she had without being any the better for it; in fact, she was getting worse.*
> *27 She had heard about Jesus, and she came up through the crowd and touched his cloak from behind, thinking,*
> *28 "If I can just touch his clothes, I shall be saved."*
> *29 And at once the source of the bleeding dried up, and she felt in herself that she was cured of her complaint.*
> *30 And at once aware of the power that had gone out from him, Jesus turned round in the crowd and said, "Who touched my clothes?"*
> *31 His disciples said to him, "You see how the crowd is pressing round you; how can you ask, 'Who touched me?' "*
> *32 But he continued to look all round to see who had done it.*
> *33 Then the woman came forward, frightened and trembling because she knew what had happened to her, and she fell at his feet and told him the whole truth.*
> *34 "My daughter," he said, "your faith has restored you to health; go in peace and be free of your complaint."*

The hemorrhaging woman says in an internal monologue that if she only but touches Jesus's clothes she may be made well. This situation, in which the person looking for healing takes the initiative, stands out as an exception. Also exceptional is the illness (vaginal bleeding) from which she suffers. Judaic purity laws consider her to be impure. That might be the reason why she approaches Jesus from behind. Haber (2003) defends a middle position between the scholars who are of the opinion that the purity laws are irrelevant to the story and those who interpret the story as a critique of Jewish purity laws. The woman's health is the primary concern of the story, and not her ritual impurity.

Jesus feels that a power goes out from him and wants to know who has touched him. The woman admits to touching Jesus, while she is trembling in fear and falling down before him. These actions of fear and falling down are signs that she recognizes what happened was a divine act (Baert, 2009). However, the act of being healed differs from magical stories. She is not healed by magic but by faith; faith has made her whole. The woman is healed by touch and by word. She is healed on both the physical and the spiritual level, which makes the theme of the text that of wholeness and holiness (Baert, 2009). It is her faith that makes her an example of unwavering belief (Kusters, 2009). This changes the miracle from the magical to the religious (Kusters, 2009).

Is Touch a Gender Issue?

In the Hebrew Scriptures, seeing and hearing are excellent ways to communicate the Torah and to get knowledge; however, this is the masculine way. Women enter Mark's gospel by having contact with Jesus through touch. It is a gender issue (Sidgwick, 2009). There is a parallel with the philosophy of Irigaray (1993), who also considers touch as gendered. However, the text differs from her opinion in the sense that in Mark the woman takes the initiative, while Irigaray mainly sees the woman as the passive partner. It is true that in Mark's vision men are linked to the Word and the Spirit, but the woman in this story is a great believer in Jesus, and that is the very reason why she touches him. Faith precedes the "miracle" and is not a consequence of it. The woman touches Jesus, so by this tactile gesture knowledge (because believing is a kind of knowledge) is transferred. She knows about his power; she knows about his divinity; and that is also what she experiences. Although not exclusive, the female epistemology is by consequence tactile (Sidgwick, 2009). By touch, women experience God's transformative power.

The story of Thomas can be called a counter story. According to Augustine (Sermons, 375), Thomas, a man, wants to touch Jesus. He has not yet recognized Jesus as the Lord and therefore needs to touch. He does not touch because he believes; on the contrary, he needs to touch in order to verify the very spatial and temporal presence of Jesus the Lord.

Does Touching Contribute to Healing?

The hemorrhaging woman is suffering from an illness that made her, in terms of that time, impure. Touching someone else makes him or her impure (Haber, 2003). According to Leviticus 15 and Numbers 5, the former states that one is only impure till the evening and makes it, by consequence, possible for the impure person to live a quasi-normal life. In Numbers 5, the person is excluded from living in his or her community. In this context, then, the hemorrhaging woman suffers from a fountain of blood leaving her. She is desperate as the life force is leaving her, for the loss of vaginal blood is linked to the loss of seed, and by consequence no new life by bearing children is possible (Haber, 2003). Also, the fountain of blood could lead to her death.

At the beginning of the narrative, the woman is nameless, which reduces her identity to her illness, but after she is healed Jesus calls her "daughter," which implies that a personal relationship has been established. This changes her identity dramatically. Touch is thus placed within the framework of a personal relationship, which comes through her faithful touch. The text states that the woman felt that she was healed and that Jesus, the healer himself, is not untouched in the process; he felt that the power has gone forth from him. However, it is noteworthy that Jesus has not become impure by her touch (Haber, 2003).

Lalleman (1998) is of the opinion that the woman is healed by mere touch and that healing by mere touch is typically Christian, although prefigured in pre-Christian Judaism. However, we have observed that her faith played an important role in the process of healing, and Mark stresses something more to discourage magical interpretation; he focuses on Christ, who interprets what happened. It is not only by touch nor only by faith but also by his Word that she is healed, that she will be whole or well. The Greek "word" means to save, to rescue and to liberate, as well as to heal and preserve (Gaiser, 2010). There is more at stake than physical healing by mere touch. It is about being healed and saved, which are linked both to the death at the cross and to the resurrection. By his Word, Jesus creates new life for those who believe. Healing comes from being in touch with Jesus. This brings us to a deeper meaning of touch; faith gives access to the power of God through Jesus, even for those who are impure. This power gives life as the woman will be able to bear children. This story and the frame story of the daughter of Jairus make it even clearer: life prevails over death.

Within religion, touch has continued to play a role. Christians touch relics, which are body parts or objects associated with saints, in the hope of being healed. For many centuries, Christians have undertaken pilgrimages to places in order to touch relics. Just as the woman in Mark's story, the pilgrims are driven by hope and faith. Makant (2014) reminds us of the often forgotten practice of anointing, which is an excellent example of the ritual of healing touch. The healing stories of Jesus and the resurrection promise that bodies that are broken physically and spiritually are made new. Makant asserts that especially in cases in which trauma has been caused by violent touch, healing should also take place by touch: the trauma being a transgression of boundaries against the will of the person, the healing taking place with the consent of the one asking for healing (Makant, 2014). This kind of touch is ritualized and differs from daily touch.

Do Not Touch Me: Forbidden Touch (Jn 20: 11–18)

The *King James Bible* is used because the text is closer to the Greek text and the first translation of *haptein*, "to touch." It translates *mè mou haptou* by "touch me not." The Jerusalem Bible translation is "do not cling to me." In the *New Oxford Annotated Bible* and *New Revised Standard Version*, it is translated "do not hold onto me." The expression is probably best known by the Latin *noli me tangere*, translated as "do not touch me." Here is the translation:

> *11 But Mary stood outside at the sepulcher weeping: and as she wept, she stooped down, and looked into the sepulcher,*
> *12 And saw two angels in white sitting, the one at the head, and the other at the feet, where the body of Jesus had lain.*
> *13 And they said unto her, Woman, why weep you? She said unto them, because they have taken away my Lord, and I know not where they have laid him.*
> *14 And when she had thus spoke, she turned back, and saw Jesus standing, and knew not that it was Jesus.*
> *15 Jesus said unto her, Woman, why do you weep? Whom do you seek? She, supposing him to be the gardener, said unto him, Sir, if you have*

> *borne him away, tell me where you have laid him, and I will take him away.*
> *16 Jesus said unto her, Mary. She turned, and said unto him, Rabbuni; which is to say, Master.*
> *17 Jesus said unto her, Touch me not; for I am not yet ascended to my Father: but go to my brethren, and say unto them, I ascend unto my Father, and your Father; and to my God, and your God.*
> *18 Mary Magdalene came and told the disciples that she had seen the Lord, and that he had spoken these things unto her.*

The text in Jn 20:17, "Touch me not," is exceptional; it is a *hapax theologoumenon* (Nancy, 2003), a paradoxical theological statement unique to this context. For, in Christianity, nothing should be untouchable, since the incarnation is a core element and the body of God is offered as food. Eating and touching are inextricably united; indeed, eating is touching with the tongue. Christianity could be considered to be the religion of touch. However, precisely at this moment between death and life, just before Jesus is going to the Father, he says that he does not want to be touched. At this very crucial moment, touch and prohibition to touch are mentioned. Would touch destroy his identity? Would it be inappropriate, or is touching his body an impossibility? Apparently, there are situations in which touching means a transgression.

Closeness and Distance: Who Is Allowed to Touch?

Mary's eyes and her capacity to see are not what bring her to recognition; rather it is the voice of Jesus who calls her by her name, "Mary." The articulation of her name, addressed to her deepest identity, leads to her recognition of the identity of the other. "Rabbuni," Mary answers, "Master." According to Baross (2001), this term colours the relationship between Mary and Jesus as one of disciple and master. Touch by the master belongs to a different order (Baross, 2001) than touch by the disciple. It becomes clear that in the double given of touching and being touched, the reciprocity in touch does not necessarily mean that there is a symmetric relation. Things do not touch a person in the same way that he or she actively touches those things. The same goes for human beings: touching and being touched are not identical givens.

The one in power can also decide not to touch and not let himself be touched. Jesus, the master, is in power because he is sent by the

Father. By refusing the touch, he withdraws from Mary, but by doing so at the same time he creates space for her. Just as he has been sent, he now sends Mary to the other disciples with the words, "I ascend unto my Father, and your Father; and to my God, and your God." That is how Mary Magdalene has become "the apostle to the apostles" (Jansen, 1998, p. 67), and consequently a new identity is created. The distance that is required by the master is thus not necessarily negative. On the contrary, the distance is necessary for the disciple to be able to become herself. It is in the withdrawal of the master, in the liberated space, that Mary can find her own place.

In Western culture the one with a higher position (the master) is culturally allowed to touch, and the one in a lower position (e.g., while he invites her to pass before him in the elevator) is not allowed to touch the one in the higher position. Having the right to touch is by consequence a sign of being in power. There are many paintings representing the scene inspired by the words *noli me tangere* in which we see that Jesus withdraws from Mary Magdalene, who wants to touch him. There exist, however, paintings in which Jesus touches her—for example, in Lucas van Leyden's engraving of Christ appearing to Magdalene (*Christ Appearing to Saint Mary Magdalen as a Gardener*, 1519, Museum Boijmans van Beuningen, Rotterdam, Netherlands).

This reality poses questions for a counsellor, pastor, or minister. When can I touch, since this is an action that only can be induced by me, the powerful one? What does it do to the other, the client, the parishioner, or the child? Is it leading to growth or to even more complications? How will my touch be interpreted by the other? And how will I touch? What kind of touch (kiss, caress), where (place), how long (duration), and with how much pressure (squeeze)? He or she must always take into account that it is the just measure of distance that creates the space for the other to find his or her own way.

The Impossibility of Touch and Transgressive Touch

The story of the hemorrhaging woman tells about inappropriate touch because the woman was impure. However, according to Nancy (1993), in the story of Mary it is not a question of inappropriate touching, but rather of the impossibility of touch. The body of Jesus is no longer part of earthly reality; it is a glorious body. Certification through touch is at the core of earthly reality, thus making Jesus's body impossible to touch.

For Irigaray (1993), however, the prohibition against touch is not a question of inappropriateness or something intangible which is impossible to touch. Rather, it is about something improper and illicit. The fact that Mary does not touch the body of Jesus means that she respected the boundaries between the human and the divine. The story refers to the Fall, where Adam and Eve did transgress the interdiction not to touch and not to eat. In Genesis 3:3 it says, "but God said, 'You shall not eat of the fruit of the tree that is in the middle of the garden, nor shall you touch it, or you shall die.'" The scene at the tomb calls to mind the passage in the creation narrative in which God forbids eating from the tree of the knowledge of good and evil. Both scenes are reminiscent of one another because they are both located in a closed garden, which is exceptional; in both, God/Christ is present in the garden. The question, "who are you looking for," echoes the question, "where are you?" Both stories contain a prohibition against touch, and in both instances this is related to life and death. The links with this text merit a more elaborate approach that, however, is not possible within this chapter. Given the links between these two passages and that Mary Magdalene did not transgress the prohibition, she has been given the name "the New Eve" (Gregory the Great, 1892, p. 189).

Derrida (2000) reminds us of the primary laws of touch that should be respected. An animal does not only die because of a lack of touch; it also dies when the touch is of such an extensive intensity that it destroys the organ of touch (Aristotle, 2001). Hence, Derrida (2000) makes the statement that a certain tact, a certain measure, is necessary: "'you will not touch,' 'you will not let others touch you,' 'you will not touch yourself too much'" (p. 61–62; free translation for the present author). The quoted laws are considered by him to be the primary laws of touching. Touching has to do with life and death. Without touching, (wo)man dies, and violent touch has the same result. Violent touch not only refers to physical violence, but also to touching without consent.

Conclusion

The texts explored in this chapter indicate that touch has to do with life and death. They refer to healing touch but also to being touched on a deeper level by Christ, who promises that life is stronger than death and that new life is always possible.

Touch always includes reciprocity that recognizes alterity. Touch also recognizes the negation of reciprocity; it gives access to knowledge that is direct and spatial. It cannot be influenced by representations. Touch brings forth a discourse on power, on closeness and distance, on both the respect of boundaries and the transgression thereof. Philosophy seems to indicate that touch is gendered and belongs to women. However, it is dangerous to think this way because this discourse can easily lead to discrimination. (Female) touch would only provide knowledge that is less important than the knowledge that (male) sight and word can bring. In fact, the *noli me tangere* has brought about the interdiction for women to touch sacred objects, to bring communion to sick people, and to preach and baptize (Jansen, 1998). This discrimination cannot rely on the biblical texts; the women in the stories have not been punished for seeking to touch. On the contrary, the hemorrhaging woman is healed, the daughter of Jairus resurrected. Mary Magdalene becomes a model of faith: the New Eve and the Apostle to the Apostles. Also, the hemorrhaging woman becomes a model of faith. On the other hand, men should not be excluded from knowledge by touch. In the stories of the appearances, Thomas was invited to touch Jesus because he doubted the reality of his presence. All men and women have access to the knowledge that all senses are able to provide. Both men and women can profit from healing touch physically and spiritually, and both are vulnerable and can be hurt by transgressive touch.

References

Augustine. (1995). 375C. *375 Sermons*. New York: New City Press.

Aristotle. (2001). Metaphysics. In R. McKeon (Ed.), *The basic works of Aristotle*. The Modern Library: New York.

Aristotle. (2001). On the Soul. In R. McKeon (Ed.), *The basic works of Aristotle* (pp. 435b 4-7). New York: The Modern Library.

Baert, B., Kusters, L., and Sidgwick, E. (2012). An issue of blood: The healing of the woman with the hemorrhage (Mark 5:24b–34; Luke 8:42b–48; Matthew 9:19–22) in Early Medieval Visual Culture. *Journal of Religion and Health, 51*/3, 663–681.

Baert, B. (2009). Who touched me and my clothes? The healing of the woman with the hemorrhage (Mark 5:24b–34parr) in Medieval Visual Culture. *Annual of the Antwerp Royal Museum* (appeared in 2011), 9–51.

Baross, Z. (2001). Noli me tangere. For Jacques Derrida. *Angelaki*, 6(2), 149–164.

Bremer, J. (2011). Aristotle on touch. *Forum Philosophicum, International Journal of Philosophy, 16*/1, 73–87.

Demasure, K. (2006). L'Épistémologie et la Théologie Pratique: Le paradigme herméneutique. In Santedi, K. and Malou Nyimi, M. (Eds.), *Epistémologie et théologie. Les enjeux du dialogue foi-science-éthique pour l'avenir de l'humanité. Mélanges en l'honneur de S.Exc. Mgr Tharcisse Tshibangu Tshishiku pour ses 70 ans d'âge et 35 ans d'épiscopat* (Recherches Africaines de Théologie 18), Kinshasa, Facultés Catholiques, 531–548.

Demasure, K. and Muller, J. (2006). Perspectives in support of the narrative turn in pastoral care. *Ned. Geref. Teologiese Tydskrif* (NGTT), 47/3 and 4, 410–419.

Democritus. (2014). Retrieved from http://www.egs.edu/library/democritus/biography/

Derrida, J. (1993). *Le toucher*, Touch/to touch him. *Paragraph*, 122–157. Jean-Luc Nancy, Paris: Bayard.

Derrida, J. (2000) Le toucher. Jean-Luc Nancy. Paris: Galilee.

English, R. B. (1915). Democritus' theory on sense perception. *Transactions and Proceedings of the American Philological Association*, *46*, 217–227.

Gaiser, F. J. (2010). In touch with Jesus: Healing in Mark 5: 21–43. *Word & World*, *30*(1), 5–158.

Gregory the Great. (1892). *De apparition Christi Magdalenae facta*. (Patrum opuscula selecta 2, hom. 25). Innsbruck: Libreria Academica Wagneriana.

Haber, S. (2003). A woman's touch: Feminist encounters with the hemorrhaging woman in Mark 5: 24–34. *Journal for the Study of the New Testament*, 171–192.

Irigaray, L. (1993). *An ethics of sexual difference*. New York: Cornell University Press.

Jansen, K. L. (1998). Maria Magdalena, in B. Mayne Kienzle and P. J. Walker (Ed.), *Women preachers and prophets through two millennia of Christianity*. Berkeley: University of California Press.

Jay, M. J. (1993). *Downcast eyes: The denigration of vision in the twentieth century French thought*. Berkeley: University of California Press.

Jonas, H. (1982). The nobility of sight: A study in the phenomenology of the senses, in *The phenomenon of life: Towards a philosophical biology*. Chicago: Harper and Row/Dell.

The Jerusalem Bible. (1966). Reader's Edition with Abridged Notes. New York: Double Day.

Kambaskovic, D. and Wolfe, C. T. (2014). *The senses in philosophy and science: From the nobility of sight to the materialism of touch*. Retrieved from http://www.academia.edu/3146325/

King James Version Bible. (1999). New York: American Bible Study.

Kusters, L. (2009). Who is she? On the identity of the hemorrhaging woman and her Wirkungsgeschichte. *Antwerp Royal Museum Annual*, 99–133.

Lalleman, P. J. (1998). Healing by a mere touch as a Christian concept. *Tyndale Bulletin*, *48*(2), 255–361, 356–357.

Laquelle, E. (ed.). (1980). *Textes pour Emmanuel Levinas*. Paris: Editions Jean-Michel Place.

Makant, M. (2014). Transforming trauma: The power of touch and the practice of anointing. *Word and World, 34*(2),164.

Mayne Kienzle, B. and Walker, P. J. (Ed.). (1998). *Women preachers and prophets through two millenia of Christianity.* Berkeley: University of California Press.

Nancy, J.-L. (2003). *Noli me tangere. Essai sur la levée du corps.* Paris: Bayard.

Nancy, J.-L. and Lacoue-Labarthe, P. (1993). Noli me Frangere. In Nancy, J-L, *The birth to presence* (pp. 266–68). Stanford, CA: Stanford University Press.

Plato. 1945. *The Republic,* Translated with Introduction and Notes by F. MacDonald. London-Oxford-New York: Cornford Oxford University Press.

Ricoeur, P. (1986). *Du texte à l'action.* Paris: Le Seuil.

Ricoeur, P. (1983). *Temps et récit, I, L'intrigue et le récit historique.* Paris: Le Seuil

Sidgwick, E. (2009). Tactility and potentiality in the motif of the hemorrhoissa. *Antwerp Royal Museum Annual,* 135–61.

Thomasset, A. (1996). Paul Ricœur. Une poétique de la morale. Aux fondements d'une éthique herméneutique et narrative dans une perspective chrétienne. *(Coll. Bibliotheca Ephemeridum Theologicarum Lovaniensium)* Leuven.

Van Leyden, L. (1519) Christ Appearing to Saint Mary Magdalene as a Gardener. Retrieved from http://collectie.boijmans.nl/en/collection/bdh-2158-pk

Vasseleu, C. (1998). Testures of light, textures of light, vision and touch. *Irigaray, Levinas and Merleau-Ponty* (Warwick Studies in European Philosophy). New York: Routledge.

Winter, A. (2001). Touching skin. Demarcating the corporeal and conceptual. AWSA Conference Proceedings. *Institute for Women's Studies.* Retrieved from http://www.ssn.flinders.edu.au/wmst/awsa2001/pdf/papers/Winter.pdf.

Wyschogrod, E. (1980). Doing before hearing: On the primacy of touch. In E. Laruelle (Ed.), *Textes pour Emmanuel Levinas* (pp. 179–203). Paris: Jean Laplace.

PART II

THE PRACTICE OF TOUCH IN PSYCHOTHERAPY

Chapter V

The Intervention of Touch in Psychotherapy and Trauma Treatment

Patricia Berendsen

Introduction to Touch in Psychotherapy and Trauma Treatment

Therapists regularly deal with psychological issues that arise from developmental disruptions and trauma-related experiences. As clinicians we are brought into the lives of our clients when they are most vulnerable, fearful, desperate, feeling alone in the world, and wounded. We find ourselves in direct contact with stories that illustrate various aspects of trauma.

Touch in therapy could be an important therapeutic tool, particularly in areas of trauma, where touch was abused or withheld. However, touch in trauma and therapy is usually hinged on one word: DON'T! More often than not, we are discouraged from touching our clients for fear of crossing boundaries, transference or counter-transference, or allegations of wrongdoing. Older (1977) and Wilson (1982) echoed sentiments that continue to this day: Many therapists have used touch but do not talk about it. Clinicians tend to be intrigued by the flood of scientific data supporting the relationship between the body and psychotherapy. However, most mainstream therapists trained in talk therapy are ill-equipped to make the transition to include somatically based modalities (Berendsen, 2011, p. 32). In fact, research by Strozier, Krizek, and Sale (2003) suggests that 82% of social workers report that touch was inadequately addressed in their educational training (p. 57).

It would appear that the clinical landscape regarding touch is changing and furthermore *needs* to change. Neuroscientists have confirmed the integration of the body, mind, and emotions. Thus, somatic approaches to therapy, including touch, will become increasingly necessary within our talk therapy framework. We will need to develop body- and touch-literacy skills. The importance of a connected and empathic therapeutic presence is vital (Berendsen, 2011; Hutterer and Liss, 2006). Touch is one of the ways that a client can experience this empathic response.

Although some relationships have had the impact of wounding and hurting a human being, other relationships, including the therapeutic one, can nurture and support healing. One byproduct of trauma can be isolation. Connection and relationship, however, can become the antidote to trauma and isolation. Touch can be a necessary intervention, which in turn can help to establish a sense of self, a sense of belonging, and connection. Touch can also be highly effective in enhancing therapeutic alliance, which is the best predictor of positive therapeutic outcomes. As such, touch in therapy may be moving from an occasional occurrence to an essential and staple intervention.

Exploring Theory and Research: A Rationale for Touch in Trauma Therapy

Touch is inherent to our humanity. Our survival and ability to thrive is dependent on touch (Montagu, 1971; Feldman and Eidelman, 2007). As infants we are touched, cuddled, and rocked, yet instances of touch decline as we grow older. This reduction in touch occurs despite the fact that our "touch hunger" does not diminish over time. From this perspective, most of us could be suffering from "skin starvation." Suffice it to say that being human and needing touch are integrally connected. An awareness of ourselves through skin contact of some sort does seem to be important for an ongoing sense of self. Touch is one of the most essential elements of human development: "The communications we transmit through touch constitute the most powerful means of establishing human relationships, the foundation of experience" (Montagu, 1986, p. xv).

Touch is important in bonding and attachment (Ainsworth and Wittig, 1969; Bowlby, 1988; Harlow 1959). Touch and sensation are our first language. According to Montagu (1971), the skin is the "self's organ of embrace and contact" (p. 205). We are bathed in uterine fluid before we are born. We travel the passage to our birth through a tight canal that massages our body, thus preparing us to awaken to our new form of existence. We are hugged and rocked and soothed as infants and throughout our childhood. These nurturing activities help us to formulate our attachment systems and our felt sense of the world being a safe place. Loving touch in the early years is essential to adequate neurological and emotional development (Bowlby, 1969; Harlow, 1971).

According to Montagu (1971), "any significant failure in the experiences of such contacts may lead to a profound failure or disorder to later interactional relationships . . . as well as in a variety of other behavioural disorders" (p. 205). If we were raised in homes where chaos, addiction, abuse and violence occurred, we are more likely to be negatively impacted by the absence of touch. Perry (2006) suggests that our attachment is our ". . . memory template for human-to-human bonds. This template serves as our primary 'world view' on human relationships" (p. 85). However, when human beings have not been satisfied by good nurturing in infancy, there remains an intense need for holding and touch (Vereshack, 1993). "Children who don't get consistent, physical affection or the chance to build loving bonds simply don't receive the patterned, repetitive stimulation necessary to properly build the systems in the brain that connect reward, pleasure and to human-to-human interactions" (Perry, 2006, p. 86). In fact, the lack of positive childhood touch has been linked to antisocial behaviour (Phelan, 2009, p. 98). A qualitative study by Steckley (2012) explored the role of touch and physical restraints in residential treatment settings in Scotland. The study found that the staff experienced some anxiety about touching residential clients. Physical restraints are to be utilized by staff as a last resort to control or subdue clients' out-of-control behaviour. So it is a sanctioned form of touching. However, for anyone who has observed a child being restrained, it is anything but pleasant and is often traumatizing for both staff and clients. Yet in Steckley's (2012) research, residential clients would use physical restraints to meet their need for touch and containment. The combination of the fears of the staff to touch and the clients' need for touch resulted in the increased use of physical restraints. Interestingly, my experience as a team leader of a children's residential treatment setting also confirms Steckley's findings.

Nuszbaum, Voss and Klauer (2014) found that participants who were briefly touched on the shoulder before entering the laboratory for a product evaluation task were more confident when they were briefly touched (p. 31). Koole, Sin, and Schneider (2014), in their study on touch and self-esteem, found that interpersonal touch alleviated existential concerns among individuals with low self-esteem. In addition Debrot, Schoebi, Perrez, and Horn (2013) found that couples who engaged in regular physical contact experienced better psychological well-being even at their six-month follow-up. This study provides evidence that intimate partners benefit from touch on a psychological level, conveying a sense of strengthened bonds between them that enhances affect and well-being (p. 1373).

The Importance of the Nervous System in Emotional Regulation and Trauma Treatment

Neuroscientists are indicating that there is a reciprocal relationship between the body and mind: "But when encountering patients with emotional problems, mental health professionals seemingly ignore the importance of the body to one's emotional stability" (Wilson, 1982, p. 65). This development of the importance of the body/mind connection is shifting therapists' perspectives and necessitating the inclusion of additional knowledge in physiology (Shaw, 1996). "What is physically lived . . . is no longer separable from neurophysiological modifications of the brain and the nervous system. The body informs the brain about sensations, communicating between the mind and the brain, and it expresses a relational affectivity with its socio-cultural context" (Andrieu, Laloe and Klein, 2012, p. 157). Touch can be utilized to re-establish regulation in the nervous system and increase the capacity for self-regulation. This can occur by specific touch interventions that can interrupt stress and threat physiology. Through these repeated interruptions, new neural pathways can be forged so that nervous system regulation can occur. Pinson (2002) suggested that therapists who used touch believed that the client's longing for touch reflected a need for attachment, self-calming and regulation.

Berendsen (2014) outlines some helpful foundational strategies when incorporating the body into trauma therapy. Her acronym of S-A-F-E-T-Y formulates a foundation for integrating the body and touch into trauma treatment.

S=Stabilize and go Slowly;

A=Attunement, Awareness and Acceptance;

F= Focus on the Felt Sense of the Feeling;

E=Empathize, Educate, Explore;

T= Take Time, Titrate;

Y= Your needs matter too! (p. 124).

(For an expanded version of S-A-F-E-T-Y, see Berendsen, 2014, p.124–127. For additional guidelines on working with trauma survivors and touch, see Benjamina and Sohen-Moe, 2005, p. 215–240.)

Nervous system dysregulation can be observed particularly if one has a lens from which to view it. For example, clients will have constriction patterns in their body that are the result of psychological and physiological responses to threat. It is important for therapists to be aware of these dimensions, as they can inform the direction of touch in therapy. Shoulders held high, darting or wide-open eyes, gastrointestinal issues, and breathing problems are some of the physical complaints that clients may exhibit along with their mental health issues. Physical constriction patterns may have become chronic and may be indicative of other relational or psychological difficulties. Typically, these patterns are a visible manifestation of a client's effort to survive. They are self-protective strategies that once were effective and now are getting in the way of optimal functioning. Touch can be used to reduce the symptoms of trauma.

Understanding where clients are in the five stages of the threat response cycle identified by Peter Levine (1997) is also helpful (Berendsen, 2011, p. 35–39). Clients may exhibit indications that they are stuck in (1) Startle/arrest/preparatory Orienting; and/or (2) Defensive Orienting Response; and/or (3) Specific Defense of Fight/flight/freeze/submit/immobility Response; and/or (4) Discharge (shaking and trembling, warmth, deep breath) and Completion; and/or (5) Exploratory Orienting Response. Accurately identifying the stages of the threat response cycle can enable increased precision in the application of any therapeutic intervention, especially touch. For example, a client in the stage of discharge/completion, which typically occurs after coming out of shock and immobility, can be regulated by physical support. The proximity of the therapist and/or touch on the upper back, shoulder or arm can be stabilizing and comforting to the client. This physical contact combined with verbal assurances can reassure the client that the shaking and trembling are normal and will subside.

Potential Impact of Touch on Clients

Touch has been shown to be therapeutic. Most touch research has been conducted in relationship to touch and massage by Tiffany Field and the Touch Research Institute. However, more research is being done on touch in psychotherapeutic settings, indicating the positive effects of touch with clients (Hunter and Struve, 1998; Strozier, Krizek, and Sale 2003; Peloquin, 1989; Horton, Clance, Sterk-Elifson and Emshoff, 1995; Salzmann-Erikson and Eriksson, 2005; Rasmark, Richt, Rudebeck, 2014). In general, clients who were touched reported an overall sense of feeling better about being in therapy (Horton, Clance and Sterk-Elifson, 1995). More recently, Dunbar (2010) has suggested that touch is linked with the release of oxytocin and endorphins, neurochemicals that support bonding and reducing our experience of pain. Positive physical aspects of touch include lifting mood in the treatment of depression (including post-natal depression), reducing anxiety, pain relief, reduction in muscle tension, decreasing raised blood pressure, enhancement of immune function, improving sleep, decreasing the symptoms of sexual abuse, reducing aggression in adolescents, and improving weight gain in preterm neonates (Field, 2003; Westland, 1993, 1993a).

Touch facilitates containment and safety (Eiden, 1998; Mintz, 1969; Hunter and Struve, 1998; Courtney and Gray, 2014). Bassya (2002) found that touch provided clients with a sense of safety, and grounded them in the present. Anxiety and dissociation can often be supported with touch, allowing a client to feel his/her body (Phelan, 2009). Close proximity with the therapist—a hand on the arm, or even the therapist touching the client's foot with his/her own—can be experienced by the client as reassuring, soothing, and grounding. Touch also serves to assist the client in focusing on the here and now (Eyckmans, 2009).

Touch is culturally specific (Jourard, 1968, p. 137). Cultural sensitivity is a must when working with clients. Communication and negotiation of any touch intervention is necessary with every culture.

Eyckmans (2009) describes in detail her experiences as a therapist in different cultures, illustrating the importance of context and utilizing a culturally sensitive framework. According to Kepner (2001), "we must come to understand how we embody the cultural, as well as personal beliefs, and attitudes that make touch forbidden or frightening" (p. 74).

Touch supports connectedness with others (Courtney and Gray, 2014; Salzmann-Erikson and Eriksson, 2005). Horton, Clance and Sterk-Elifson (1995) found that clients indicated a deeper trust and stronger attachment with the therapist when touch was used in therapy. Clients attribute touch to creating a feeling of bond, closeness or a sense that the therapist really cares. Salzmann-Erikson and Eriksson (2005), in their research on the meaning of touch with patients who have been treated for psychosis, describe that through touch, clients can feel a sense of belonging and kinship. In my therapy practice, I can recall a middle-aged woman with severe eating issues. When I responded warmly to her request for a hug at the end of her session she looked up with tears in her eyes and said, "You know, I can't believe you would hug me. I didn't know I was deserving of being touched by anyone. I don't feel so unlovable. Thank you!"

Touch facilitates communication and elicits comfort. Clients tend toward increased self-disclosure when touched (Pattison, 1973; Mintz, 1969; Eiden, 1998). Salzmann-Erikson and Eriksson (2005) reported that according to clients "feelings get communicated in the act of touching" (p. 848). In addition, touch contributes to greater openness with the therapist. As clients are able to verbalize their feelings, they increase their potential to reach developmental milestones that may have been missed (Goodman and Teicher, 1988; Tune, 2005). Montagu (1971) asserts that "taking almost anyone's hand under conditions of stress is likely to exert a soothing effect, and by reducing anxiety and giving a feeling of greater security" (p. 216). "To give someone your hand is not just a physical action. It is a welcome into a shared world that therapists, but also clients initiate" (Rasmark, Richt, and Rudebeck, 2014, p. 5). Salzmann-Erikson and Eriksson (2005) indicated that "The need for touching becomes stronger when one's mental health is in deterioration. Touch from another human being has a comforting and supportive function" (p. 847).

Touch promotes awareness and affirms the sense of the self. Touch can facilitate and support the here-and-now experience of the client. This can allow the client to be with an intense feeling or emotion and keep their "witness" on board. In this way, touch can help a client to maintain dual awareness of both the present and the past. In addition, the client's self-awareness tends to precede increased self-exploration (Courtney and Gray, 2014). Touch can support the acquisition of skills to enable clients to identify and experience their physiological sensations (Davis, 2001). Touch also affirms the sense of self (Peloquin, 1989; Mintz, 1969; Eiden, 1998; Courtney and Gray, 2014). Through touch, a therapist can communicate the message that "I accept you," "I see you," or "I am here for you." Horton, Clane, Sterk-Elifson and Emshoff (1995) examined individuals in therapy with a non-body-oriented psychotherapist who experienced some sort of physical contact beyond accidental contact or a formal handshake with the therapist. Interestingly, 71% of patients who reported a history of abuse identified that touch enhanced self-esteem, trust, and a sense of their own power. Some patients reported that touch helped them to feel that they were worthy of respectful touch, stating, "Touch helped me learn I was lovable" (p. 451).

Touch assists in the development of intimacy. Rasmark, Richt and Rudebeck (2014) noted that professionalism involves using not only our competencies, but also our physical and emotional contact to deepen the relation. When people touch, the distance between them decreases. Touch can be a means of closeness in the therapy room. However, since touch is a form on intimacy, the timing of touch is important. Nuances of knowing when to touch and when not to touch deepen the therapeutic relationship.

Touch helps client access/process pre-verbal material (Shaw, 1996). Strozier, Krizek and Sale (2003) suggest that touch itself may facilitate clients "getting in touch" with emotions that may not be obvious in psychotherapy. "Respectful, reassuring touch seemed to help many patients feel supported and safe enough to move into threatening material on a deeper level" (Horton et al., 1995, p. 451). Furthermore, touch can be a means of processing pre-verbal material that literally has no voice, only sensations. In addition, touch can facilitate symbolic parenting when the client is incapable of verbal communication, perhaps where there has been a deficit in childhood (Bosanquet, 1970; Mintz, 1969; Toronto, 2006). In these instances, it is imperative for the therapist to meet the client at the somatic level as words can sometimes be experienced as misattunement.

Touch is needed when someone is ill. Touch can bring us out of a distressed state with considerable ease. Bowlby writes that the need for touch increases with danger, incapacity and sickness (as quoted in LeMay, 1986, p. 28). Illness, according to Jourard (1964), occurs when "a person's life begins to lose zest, a sense of future, meaning, and love" (p. 138). This is confirmed by several other researchers (Benjamin and Sohnen-Moe, 2005; Horton, Clance, Sterk-Elifson and Emshoff, 1995; Huss, 1977; Mintz, 1969; Strozier, Krizek and Sale, 2003).

Touch increases a sense of empowerment and aliveness. For many clients their sense of power in defining and defending boundaries has been thwarted in the wake of trauma. Touch may have been imposed upon them. Saying no to the therapist's invitation for touch support without any negative repercussions (punishment or rejection) can be reparative and liberating for the client. Likewise, open dialogue about touch between client and therapist supports the client having a sense of control (Geib, 1998). Touch increases aliveness (Eiden, 1998; Jourard, 1968; Levine 2010, 1997), ". . . the experience of being touched enlivens our bodies, and brings us back into them" (Jourard, 1968, p. 148). Even in intimate partner relationships, touch promoted and strengthened bonds between partners. Participants who were touched more during the study reported better psychological well-being six months later (Debrot, Schoebi, Perrez, and Horn, 2013, p. 1373).

Types of Touch in Psychotherapy

Most codes of ethics do not focus on nonsexual physical contact between counsellor and client (Calmes, Piazza and Laux, 2013). Typically, the forms of touch that are most commonly used by therapists and that generally do not constitute misconduct are handshakes, touching the shoulder, arm or hand of the client, hugging, and holding hands (Stake and Oliver, 1991; Strozier, Krizek, and Sale, 2003). Hugging, holding a client in distress, stroking or patting, and a kiss on the cheek were also mentioned (Tune, 2001). Jourard and Rubin (1968) found that the most acceptable parts of the body to touch are the hand, lower and upper back, shoulder region, and middle back. Sexual or violent touch is considered unethical in every discipline.

Not touching is also an intervention. This can potentially be felt by a client as rejection and could minimize the opportunity for growth (Salzmann-Erikson and Eriksson, 2005; Wilson, 1982; Zur, 2007). On the flip side, a client's request for physical contact declined by the therapist might be a new experience of boundaries and personal agency. This may be a new experience for a client of setting limits and/or asking for what they need or want. Respectful touch contains within it the possibility of communicating appropriate boundaries. Another component that is paramount is that therapists who are not comfortable using touch make it explicit to their clients. In this way, therapists can inform their clients of their therapeutic stance so that clients are not "shamed by the need for physical reassurance or comforting" (Horton, Clance, Sterk-Elifson and Emshoff, 1995, p. 455).

The importance of knowing how and when to touch, as well as where to touch and for how long, are elements that a therapist needs to consider. "When physical contact was unwanted or undesired, it gave rise to feelings of inferiority, fear, and annihilation, which were experienced as suffocating and oppressive" (Salzmann-Erikson and Eriksson, 2005, p. 849). The challenge for therapists is that most do not have much training in the specifics of touch in their therapeutic work (Caldwell, 2000; Tune, 2001). Milakovich (1993) found that therapists who used touch were less concerned about risk and tended to use their own judgment about when to use touch.

Westland (2011) proposed that touching in the therapeutic relationship requires "experiential training." This includes knowledge of how to touch contactfully, having a coherent theoretical perspective, knowing how to monitor both one's own and the client's responses, being able to discuss touch as an aspect of the ongoing relationship, and having adequate supervision from someone who has also had touch training (p. 27).

"Only when the therapist has learned the ins and outs of his or her own sensations and emotions and is relatively comfortable with them, can the therapist really help clients contain their own troubling sensations and emotions so that they can learn that, no matter how horrible they feel, it will not go on forever" (Levine, 2010, p. 47). Our somatic responses to our clients can tell us something about them *and* about us. When we transform our responses, we become co-regulators of the transformation of our clients' experience, too! Becoming aware of our body sensations prior to, during, and after sessions can help monitor our own regulation or dysregulation. Vanderheyden, as quoted in Berendsen (2006), echoes this sentiment by suggesting that, "not being reflective practitioners makes us liabilities to our profession" (p. 17).

Trainings that utilize touch and the body more globally in psychotherapy are Somatic Experiencing, Sensorymotor Psychotherapy, Bodynamic Analysis, Emotionally Focused Therapy, Psychobiological

Approach to Couples Therapy, EMDR, and Gestalt therapy, to name only a few. Given that most therapists have little or no training or exposure to touch in their therapy, it would be helpful for therapists to expand their skills by taking such courses.

Guidelines for Safe and Effective Touch in Psychotherapy

Dr. Seuss (1993) states, "so be sure when you step, step with care and great tact, and remember that life is a great balancing act." This is especially true when integrating touch in therapy. Effectively utilizing touch means being flexibly responsive to the client who is sitting in front of us. Westland (2011) suggests that touch is informed by presence, intentionality, and congruence. "Contactful touch," she says, "happens in the here and now, moment by moment" (p. 27). Although this is true, there are some basic elements that contribute to safe and effective touch.

Permission, permission, permission. Dialogue is critical when a therapist uses touch. It is essential for the clinician to offer touch with the caveat that the client is welcome and encouraged to say yes or no to any proposed touch intervention. Following this, it is important that each touch intervention be followed up with further permission from the client. Asking permission at several points during any touch intervention communicates clearly the therapist's intention that the client's needs are the priority.

Attunement. The onus is on the therapist to develop their attunement skills, which are vital to utilizing touch. Being able to accurately read verbal and non-verbal cues as well as having dialogue that is open and ongoing with clients about touch is key. As Westland (2011) states, "The client and the therapist *at this moment and with this client* should be comfortable with touch" (p. 28). Attunement is a skill that can be developed. This can be accomplished as clinicians engage in their own personal therapeutic work and deepen their understanding of developmental needs.

Intention precedes touch. "Intention is energy and may be experienced before the actual physical touch occurs. Intentionality is fundamental to relating; and the intention of the giver makes a difference to the touch, and how it is received. In therapeutic work the client and therapist co-create the field together" (Westland, 2011, p. 18). Carlsson (2003, referred to in Salzmann et al., 2005) shared the perspective that touch that is nurturing or not nurturing depends on the underlying intention of the touch. The therapist's intention supports clarity when utilizing touch. There is less confusion for clients when the touch is congruent and there is clarity regarding boundaries. In this way, the client experiences the felt sense of the touch being for his or her benefit versus for the needs of the therapist (Geib, 1982; Horton et al., 1995).

Touching needs to be collaborative. Working together with the client is critical when using touch in therapy. Checking in with the client from moment to moment can help the client to feel involved and supported in the therapy process. "There is a sense of mystery. No helper can presume to understand fully. There is respect for what cannot be known, what cannot be understood. This profound respect, fused with an equal belief in the potential for helping, shapes a view of the person being helped as that of a vital collaborator" (Jourard, 1968, p. 25). Gelb's (1982) research indicated that many patients have difficulty requesting physical contact or sharing their negative reactions about therapy. Thus, having frank conversations and negotiating aspects of touch are essential when doing any form of touch work.

Therapist commitment to self-awareness is paramount. "If we dare to do therapy with people, we had better be aware of our own internal process, drives, needs and styles of relating to others. Unless we have dealt with our own issues, we may be tempted to focus on our own unmet needs [for power and control]" (Collins, 1987, p. 208). Kepner (2001) underscores the need for therapist self-awareness, stating, "The understanding of one's self and biases is a prerequisite for any therapeutic application, but is even more essential for such intimate and directly contactful work as touch" (p. 74). Vanderheyden (Berendsen) (2006) echoes this in her prose "I Dare You," written from the perspective of a client speaking to the therapist, saying, "I want to know if you have enough power in your own life so you don't need to have power over mine" (p. 45.)

The therapist needs to be attentive to his/her own body (Rasmark, Richt, and Rudebeck, 2014). Courtney and Gray (2014) suggest that it is important for therapists to "understand their own attachment experiences of touch." Their findings also suggest that "sorting through one's own personal issues or countertransference" is pertinent (Courtney and Gray, 2014, p. 126). Ryan (1999) highlights that "a therapist who is not sensitive to, knowledgeable about, and accepting of transference and countertransference processes can fail at the therapeutic task and even retraumatize a client" (p. 472).

Therapist understanding of trauma physiology and the nervous system are essential. The therapist's knowledge of the threat response cycle can aid in identifying physiological states of the client. Understanding the autonomic nervous system assists the therapist in knowing what branch of the nervous system (sympathetic or parasympathetic) is active by observing the physical changes in the client. In addition, information about the polyvagal theory (Porges, 2011) will inform the clinician about the ventral and dorsal systems, which are deeply impacted by trauma. The therapist can then discern what forms of touch might be helpful or when touch may be beneficial. This information becomes vital in developing an overarching clinical formulation, as well as next steps for touch in particular.

The therapist needs clinical supervision. Courtney and Gray (2014) emphasize the importance of using "a structured training aimed at helping the practitioner to appropriately integrate touch into practice" (p. 126). Unfortunately, in the training and education of therapists, touch is a neglected aspect (Caldwell, 2000). In addition, Wilson (1982) discovered that clinicians were deterred from admitting using touch in their practice. Clinical supervision with supervisors who have had experience incorporating touch into their practice would be ideal. However, since this is not commonplace, engaging in supervision will support the therapist in the awareness of transference and countertransference. Supervision will also assist in identifying any blind spots and to ensure that the touch used in therapy has a clear rationale. Engaging in supervision can minimize potential pitfalls and strengthen ethical therapeutic practice. "Not being reflective practitioners makes us liabilities to our profession" (Vanderheyden (Berendsen), 2006, p. 17).

Case Example of Using Touch/Physical Contact with a Teenage Client

Julie, a 13-year-old, was residing in a girls' group home. Her biological mother suffered from severe substance abuse and, consequently, when Julie was eight years old, the Children's Aid Society removed her and her younger sibling from their mother's care and placed them in a foster home. This arrangement worked for awhile. However, as Julie grew, so did her temper and her depression. This proved to be too much for her foster mother to handle. Unfortunately, the placement broke down and Julie ended up living in the group home. Her younger sibling remained in the foster home.

I was the therapist serving this group home and was having my usual weekly meeting with Julie. We had been working together for about five months at this point. Therapy focused on stabilization, safety and containment, adjustment to group home living, and coping skills. However, this particular Monday, Julie was *very* upset. Her mother had been scheduled to have a visitation with Julie and her sister on the previous weekend. It did not go well at all. Julie's mother showed up two hours late. When she did show up she was intoxicated, and the visit was terminated by the social worker.

Julie sat in front of me, looking completely deflated. Her eyes were lifeless and downcast and her shoulders slumped. I had seen Julie in a discouraged state before, but never had I sensed her hopelessness as I did this day. [*Attunement. Being able to accurately read verbal and non-verbal cues.*] I reflected to Julie what I was observing and let her know that I could see her distress. I also let her know that it looked like she might be feeling hopeless. At this point her eyes made brief contact with me and then turned toward the floor again. [*Attunement, creating connection to self and others.*] I encouraged her to take her time to speak (or not). [*Attunement, collaboration and permission.*] I let her know that I felt badly seeing her suffer. I suggested that I really did not know what to do except to be with her.

We sat in silence for a while until Julie was able to and chose to speak. [*Understanding trauma physiology and the nervous system, as Julie needed time to settle and perhaps formulate words for her experience.*]. She began with, "It doesn't matter . . . why do I care . . . I always get disappointed . . . I'm so mad at myself for hoping that things will be different with my mom . . . that she will do what she says and fucking show up . . . and show up sober! If she doesn't care about me why should I care about me! I'm not worth the time." At this point, Julie began to sob. Perhaps wailing would be more accurate. I asked Julie if it would be okay to move a little closer to her. [*Permission and proximity facilitating containment and safety.*] She nodded yes. [*Touch helps the client access/process pre-verbal material—the likelihood of this client remembering all the times when she was disappointed by her mother was pretty high.*]

After a few minutes of proximity and more tears, Julie went on to describe how hopeful and excited she was to see her mother. [*Touch increases disclosure.*] She had made something in art class for her mom and was looking forward to giving it to her. As she kept waiting for her mother she reported that she became more upset, sad and confused. She shared that she was worried her mother would not come.

She talked about feeling lonely and alone, which worsened as minutes and hours passed by. In all of her upset, she ripped up her artwork.

I could observe that Julie was a bit more settled by sharing her frustration and anger and being somewhat comforted by closer proximity. [*Touch eliciting comfort and facilitating communication.*] She didn't seem as agitated and was making more eye contact, and her posture was not as slumped. Her breathing was smoother and deeper, and colour was coming back into her face. [*Touch increases aliveness.*]

I had an idea which came into my awareness that I thought might help Julie. [*Therapist commitment to self-awareness and being attentive to her own body.*] I imagined me sitting beside her, parallel, on the armrest of the couch she was sitting on, and having her slowly lean into me. I posited that it could be an interesting experiment to actually feel what it is like to lean on somebody. I wondered if being able to lean into someone who was there might give her a different experience of being met. The thought remained persistent, so I told Julie about what I was thinking. [*Intention precedes touch.*] I inquired as to whether Julie would like to try out this idea as an experiment (or not). [*Collaborative.*] I assured her that it was quite okay if she didn't want to try anything right now. [*Permission seeking and collaborative.*] I also suggested that she could take her time and think about it before she gave me an answer. [*Facilitating containment and safety.*] Julie replied, "Sure."

I moved to the armrest and sat beside her. After I did this, I invited Julie to notice what she became aware of in her body or her thinking. Julie shared that she didn't seem to notice much. This indicated to me that I needed to give Julie more time. [*Attunement and understanding of trauma physiology and nervous system. Closer proximity might need time to settle and get used to the transition.*] It was important to let her nervous system settle with the activation of someone being closer to her. When I sensed that Julie was settled, [*therapist attentive to her own body*] I encouraged her when she was ready to begin to lean into me. I assured her that I would just be there for her to lean into. I communicated that my intention was for her to perhaps have an experience of leaning on someone and to see what it was like for her. [*Intention precedes touch.*]

Julie began to lean into me very slightly at first. As she was doing so, we were exploring what her felt sense was of this experience. Sometimes there was just silence. I noticed that Julie's body seemed stiff at first. Gradually, I sensed increased softening and relaxation. I felt the increasing weight of her body leaning into me. At this point, Julie began to cry . . . softer tears coming from her eyes this time. [*Touch promotes comfort, awareness, and sense of self.*] "I don't think I've ever felt this before." "Felt what?" I replied. "I feel myself relaxing more. I didn't realize I was so tense . . . that I held so much in. I feel like I can breathe and that my stomach isn't in knots . . . it's like I am having a nap sort of . . .. " I encouraged Julie to lean into me as long as she wanted, that she had all the time she needed to soak up this new experience in her body. [*Permission and touch eliciting comfort and awareness.*] Eventually, Julie began to pull herself away from me and sat upright on the couch. [*Touch increasing empowerment as Julie was taking initiative to pull away.*] I took that as my cue that I perhaps could move away back to my chair. I checked in with Julie [*Permission*], who indicated that she wanted me to stay where I was. [*Touch increasing empowerment as Julie was asking for what she needed.*] I complied. [*Collaboration.*] We talked further, exploring this new sensation/experience of leaning on someone. I could see from Julie's eyes that this was a truly new experience for her, an "aha" moment of sorts. Her eyes seemed more alive and brighter, with the sadness still there but more in the background. [*Touch increases aliveness.*] She suggested that I could go and sit in my chair. [*Touch increasing a sense of empowerment.*] I was struck by the softness of Julie's face. [*Therapist understanding of trauma physiology.*] The quality of her eye contact with me was so much more engaged. [*Touch supports connection with others.*] It appeared that she was more energized. [*Touch promoting aliveness.*] Julie indicated to me that she was okay now and wanted to end our session. [*Touch increases a sense of empowerment.*] So we did.

Approximately 30 minutes later, group home staff approached me asking, "What did you do with Julie?" I was perplexed by their curiosity and inquired why they were asking me this question. A staff replied, "Almost immediately after your session with her, Julie came to us with a bag of broken glass. She told us that she was planning to use the glass to cut herself and maybe even try to end her life, if she cut herself badly enough. Instead, she handed in the glass voluntarily stating that she "didn't need it anymore." I was stunned and grateful at the same time. Julie's sudden change gave me an opportunity to share with the group-home staff team about the power of touch and nervous system regulation. That was a turning point for all of us. We saw and experienced the potency of a seemingly small intervention that unknowingly yielded life-saving results.

Conclusion

Touch can be an important and powerful intervention for trauma treatment. Van de Kolk (2014) suggests that "learning to experience and tolerate deep emotions is essential for recovery from trauma" (p. 6449). He also asserts that "competence is the best defense against the helplessness of trauma" (p. 6402). One of the ways that clients can learn to tolerate intense emotions and gain competence as a means of dealing with trauma is through touch. As evidenced by Koole, Sin and Schneider (2014), interpersonal touch alleviated existential concerns among individuals with low self-esteem (p. 30). Touch that is sensitively utilized by a skilled and well-trained clinician can assist in self-regulation, an absent characteristic of traumatized clients. Adhering to S-A-F-E-T-Y principles and acquiring knowledge about nervous system regulation as well as the threat response cycle supports the foundation for integrating touch into therapy. Additionally, understanding the responsibilities of the therapist and the guidelines for safe and effective touch therapy promotes ethical practice.

Aging-mentor Morrie Schwartz, in the poignant movie *Tuesdays with Morrie*, tells Mitch Albom, "In the beginning of life, when we are infants, we need others to survive, right? And at the end of life, when you get like me, you need others to survive, right?" His voice dropped to a whisper. "But here's the secret: in between, we need others as well." Touch is one of the ways that we need each other.

References

Ainsworth, M. D. S. and Wittig, B. A. (1969). Attachment and the exploratory behaviour of one-year-olds in a strange situation. In B. M. Foss (Ed.), *Determinants of infant behavior* (Vol. 4, pp. 113–136). London: Methuen.

Andrieu, B.; Laloe, A-F. and Klein, A. (2012). Touch, skin cultures and the space of medicine: the birth of biosubjective care. *Touching space, placing touch*, M. Dodge (Ed.). Aldershot: Ashgate.

Bassya, P. (2002). Touch in therapy: An effort to make the unknown known. *Journal of Contemporary Psychotherapy, 32,* 179–196.

Benjamin, B. and Sohen-Moe, C. (2005). *The ethics of touch*. Tuscon, AZ: SMA Inc.

Berendsen, P. (2014). Supporting the integration of the body into psychotherapy and trauma treatment. In T. O'Connor, K. Lund and P. Berendsen (Eds.), *Psychotherapy: Cure of the soul.* Waterloo: Waterloo Lutheran Seminary.

Berendsen, P. (2011). When words are not enough: Incorporating the body into a trauma framework and trauma therapy. *Child and Family Professional, 14*(3), 32–51.

Bosanquet, C. (1970). Getting in touch. *Journal of Analytical Psychology*. Vol. 15 (1) 42–57.

Bowlby, J. (1988). *A secure base: Parent-child attachment and healthy human development*. New York: Basic Books.

Bowlby, J. (1952). Maternal Care and Mental Health. Geneva: World Health Organization.

Caldwell, C. (2000, June). Using touch in psychotherapy. Paper presented at the USA Body Psychotherapy Conference, Baltimore.

Caldwell, C. (1996). *Getting our bodies back: Recovery, healing, and transformation through body-centered psychotherapy*. Boston: Shambhala.

Calmes, S. A., Piazza, N. J. and Laux, J. M. (2013). The use of touch in counselling: An ethical decision-making model. *Counselling and Values*. April, Vol. 58. doi: 10.1002/j.2161-007X.2013.00025.x

Carlsson, G., Dahlberg, K. and Drew, N. (2000). Encountering violence and aggression in mental health nursing: A phenomenological study of tacit caring knowledge. *Issues in Mental Health Nursing, 21*(5), 533–545.

Collins, W. (1987). Keeping the therapist alive. *Journal of Religion and Health*, Vol. 26, No. 3 (Fall, 1987), 206–213.

Courtney, J. A. and Gray, S. W. (2014). A phenomenological inquiry into practitioner experiences of developmental play therapy: Implications for training in touch. *International Journal of Play Therapy, 23*(2), 114–129.

Davis, W. (2001). Energetics and therapeutic touch. In Heller, M. (Ed.), *The flesh of the soul: The Body we work with. Selected papers of the 7th Congress of the European Association of Body Psychotherapy 2-6 September 1999, Travemünde* (pp. 59–80). Bern: Peter Lang.

Debrot A., Schoebi, D., Perrez, M. and Horn, A. B. (2013). Touch as a mediating role of psychological intimacy. *Personality and Social Psychology Bulletin, 39*(10), 1373–1385.

Dunbar, R. I. (2010). The social role of touch in humans and primates: Behavioural function and neurobiological mechanisms. *Neuroscience and Biobehavioral Reviews, 34*, 260–268.

Eiden, B. (1998). The use of touch in psychotherapy. *Self and Society. International Journal for Humanistic Psychology*.

Eyckmans, S. (2009). Handle with care: Touch as a therapeutic tool. *Gestalt Journal of Australia and New Zealand, 6*, 40–53.

Feldman, R. and Eidelman, A. I. (2007). Maternal postpartum behavior and the emergence of infant-mother and infant-father synchrony in preterm and full-term infants: The role of neonatal vagal tone. *Developmental Psychobiology, 49*, 290–302.

Field, T. (2003). *Touch*. Cambridge, MA: MIT Press.

Field, T. (1998). Massage therapy effects. *American Psychologist, 53*, 1270–1281.

Field, T., Diego, M., and Hernandez-Reif, M. (2007). Massage therapy research. *Developmental Review, 27*, 75–89.

Geib, P. (1998). The experience of nonerotic physical contact in traditional psychotherapy. In E. W. L. Smith, P. R. Clance and S. Imes (Eds.), *Touch in psychotherapy: Theory, research, and practice* (pp. 109–126). New York: The Guilford Press.

Goodman, M. and Teicher, A. (1988). To touch or not to touch. *Psychotherapy, 25*, 492–500.

Harlow, H. F., and Suomi, Stephen J. (1971). Social Recovery by Isolation-Reared Monkeys. *Proceedings of the National Academy of Science of the United States of America, 68*(7), 1534–1538.

Harlow, H. F. (1959, June). Love in infant monkeys. *Scientific American, 200*, 68–74.

Herman, J. (1992). *Trauma and recovery*. New York: Basic Books.

Horton, J., Clance, P., Sterk-Elifson, C., and Emshoff, J. (1995). Touch in psychotherapy: A survey of patients' experiences. *Psychotherapy: Theory, Research, Practice, Training, 32*(3), 443–457. doi:10.1037/0033-3204.32.3.443

Hunter, M. and Struve, J. (1998). *The ethical use of touch in psychotherapy*. London: Sage Publications.

Huss, J. (1977). Touch with care or a caring touch? *American Journal of Occupational Therapy, 31*, 1–12.

Hutterer, J. and Liss, M. (2006). Cognitive development, memory, trauma, treatment: An integration of psychoanalytic and behavioral concepts in light of current neuroscience research. *Journal of the American Academy of Psychoanalysis and Dynamic Psychiatry, 34*, 287–302.

Jourard, S. (1968). *Disclosing man to himself*. New York: D. Van Nostrand Company.

Jourard, S. (1961). *The transparent self: Self disclosure and well-being*. New York: Van Nostrand Reinhold Company.

Kain, K. L. (2012). *Touch skills training for trauma therapists*. Saskatoon, SK. Day 1 of 5, June 21, 2012.

Kepner, J. (2001). *Body process, working with the body in psychotherapy*. Cambridge MA: Gestalt Press.

Koole, Sin and Schneider (2014). Embodied terror management: Interpersonal touch alleviates existential concerns among individuals with low self-esteem. *Psychological Science, 14*(25), 30–37. doi: 10.1177/0956797613483478

LeMay, A. (1986). The human connection. *Nursing Times*, Vol. *82*(47), 28–32.

Levine, P. A. (1997). *Waking the tiger: Healing trauma*. Berkeley, CA: North Atlantic Books.

Levine, P. A. (2010). *In An Unspoken Voice: How the body releases trauma and restores goodness*. Berkeley, CA: North Atlantic Books.

Milakovich, J. C. (1993). Touch in psychotherapy: The difference between therapists who touch and those who don't. *Dissertation Abstract International, 54*, 3347.

Mintz, E. (1969). On the rationale of touch in psychotherapy. *Psychotherapy: Theory, Practice, 6*, 232–234.

Montagu, A. (1971 / 1986). *Touching: The human significance of the skin.* New York: Columbia University Press.

Nuszbaum, Voss and Klauer (2014). Assessing individual differences in the need for interpersonal touch and need for touch. *Social Psychology, 45*(1), 31–40. doi: 10.1027/1864-9335/a000157

Older, J. (1977). Four taboos that may limit the success of psychotherapy. *Psychiatry*, 40:197–980.

Pattison, J. (1973). Effects of touch on self-exploration and the therapeutic relationship. *Journal of Consulting and Clinical Psychology, 40*, 170–175.

Peloquin, S. (1989). Helping through touch: The embodiment of caring, *Journal of Religion and Health, 28(4)*, Winter, 299–322.

Perry, B. and Szalavitz, M. (2006). *The boy who was raised as a dog.* New York: Basic Books.

Phelan, J. E. (2009). Exploring the use of touch in the psychotherapeutic setting: A phenomenological review. *Psychotherapy: Theory, Research, Practice, Training, 46*(1), 97–111. doi:10.1037/a0014751

Pinson, B. (2002). Touch in therapy: An effort to make the unknown known. *Journal of Contemporary Psychotherapy, 32*, 179–196.

Porges, S. (2011). *The polyvagal theory*. New York: W.W. Norton and Company.

Porges, S. W. (2001). The polyvagal theory: Phylogenetic substrates of a social nervous system. *International Journal of Psychophysiology, 42*(2), 123–146.

Porges, S. W. (2004, May). Neuroception: A Subconscious system for detecting threats and safety. *Zero to Three*, 19–24.

Rasmark, G., Richt B. and Rudebeck, C. (2014). Touch and relate: body experience among staff in habilitation services. *International Journal of Qualitative Studies on Health & Well-Being, 9*, 1–12. doi:10.3402/qhw.v9.21901

Ryan, K. (1999). Self-help for the helpers: Preventing vicarious traumatization. In N. B. Webb (Ed.), *Play therapy with children in crisis* (2nd ed.) (pp. 471–491). New York: Guilford Press.

Salzmann-Erikson, M. and Eriksson, H. (2005). Encountering touch: A path or affinity in psychiatric care. *Issues in Mental Health Nursing, 26*, 843–852.

Shaw, R. (1996). Towards an understanding of the psychodynamic processes of the body. *Psychodynamic Counselling, 2*, 230–245.

Stake, J. E., and Oliver, J. (1991). Sexual contact and touching between therapist and client: A survey of psychologists' attitudes and behavior. *Professional Psychology: Research and Practice, 22*, 297–307.

Steckley, L. (2012). Touch, physical restraint and therapeutic containment in residential child care. *British Journal of Social Work, 42*, 537–555. doi:10.1093/bjsw/bcr069

Strozier, A. L., Krizek, C. and Sale, K. (2003). Touch: Its use in psychotherapy. *Journal of Social Work Practice, 17*, 49–62.

Suess, Dr. (1993). *Oh, the places you'll go!* New York: Random House.

Szalavitz, M. and Perry, B. D. (2010). *Born for love: Why empathy is essential-and endangered.* New York: HarperCollins.

Toronto, E. L. K. (2006). A clinician's response to physical touch in the psychoanalytic setting. *International Journal of Psychotherapy,* Vol. 7, No. 1. 69–81.

Tuesdays with Morrie [Motion picture on DVD]. (2000). Burbank, CA: Touchstone Home Video.

Tune, D. (2001). Is touch a valid therapeutic intervention? Early returns from a qualitative study of therapists' views. *Counselling and Psychotherapy Research, 1*(3), 167–171.

Vanderheyden (Berendsen), P. (2006). You never know…The power of intentional/purposeful interactions. *Child and Family Journal, 8*(3), 15–17.

van der Kolk, Bessel (2014). *The body keeps the score: Brain, mind, and body in the healing of trauma.* Penguin Group, US. Kindle Edition, 456–457.

Vereshack, P. (1993) *The Psychotherapy of the Deepest Self.* Ontario: Life Perspectives.

Wardell, D. W., and Engebretson, J. (2001). Biological correlates of Reiki touch healing. *Journal of Advanced Nursing, 33,* 439–445.

Westland, G. (1993). Massage as a therapeutic tool, part 1. *British Journal of Occupational Therapy,* April, *56*(40).

Westland, G. (1993a). Massage as a therapeutic tool, part 2. *British Journal of Occupational Therapy,* May 1993, *56*(5).

Westland, G. (2011). Physical touch in psychotherapy: Why are we not touching more? *Body, Movement and Dance in Psychotherapy: An International Journal for Theory, Research and Practice* (6) 1, 17–29.

Wilson, J. M. (1982). The value of touch in psychotherapy. *American Journal of Orthopsychiatry, 52*(1):65 72.

Zur, O. (2007). *Boundaries in Psychotherapy: Ethical and clinical considerations.* Washington, DC: APA.

Chapter VI

A Puppy's Touch: Destressing with Dogs in a University Environment

Kristine Lund

There is no psychiatrist in the world like a puppy licking your face.

– Bern Williams

Introduction

Humans interacting or having relationships with animals is not a new phenomenon. In fact, Messent and Serpell (1981) note that there is fossil evidence from half a million years ago that indicates a connection between *Homo erectus* and a canine-like species. Davis and Vall (1978) argue that an affectionate relationship between humans and dogs existed thousands of years ago. Their assertion is based on the discovery of a 12,000-year-old tomb in modern Israel where a person was buried with one arm around a puppy. Dogs have been bred to coexist with their human counterparts. They have filled many roles such as herding, guarding, hunting, fishing, and being our best friend (Clutton-Brock, 1995).

Today the relationship between humans and animals in Canada is very strong. According to Consumer Corner, Canadian Pet Market Outlook (2014), 57% of Canadian households own a pet. This translates to approximately 7.5 million households. Thirty-seven percent of Canadian households own one or more cats, while 32% own dogs. Overall, Canada is home to approximately 5.9

million dogs and 7.9 million cats. Nine percent of Canadians owned other types of pets, including fish, birds, small mammals and reptiles. CBC News (March 2014) reported that "Canadian pet owners are spending more money today than ever before to ensure their furry friends are healthy and happy." While some might question the time, energy, and money that are spent on pets in Canada, many pet owners report that their companion animals provide a great deal in the form of affectionate attachments. Archer (1977) observed the perceived mutual affection between companion animals and their human counterparts and reported the loving and pleasant feelings often experienced during these interactions. These positive interactions have spurred a unique field of research called human-animal interactions.

To understand the various types of animal-assisted interventions (AAI) the American Veterinary Association classification is used to identify three categories: animal-assisted activities (AAA) that utilize companion animals; animal-assisted therapies (AAT) that utilize therapy animals; and service animal programs (SAP) that utilize a service animal.

Animal-assisted activities (AAA) and animal-assisted therapies (AAT) have included the use of a variety of animals, such as dogs, cats, horses, birds, fish, and lizards, and have occurred in a number of different clinical contexts, such as schools, hospitals, elder care facilities, counselling settings, and group homes. Depending on the context, the animal may visit the facility with a handler or may be resident. Service animal programs typically use dogs, which reside with the individual needing this resource.

Using horses in psychotherapeutic contexts falls under a different jurisdiction, particularly the North American Riding for the Handicapped Association (NARHA) and the Equine Facilitated Mental Health Association (EFMHA).

Animal-assisted therapies are beginning to be recognized as a treatment modality similar to dance, art, and music; however, Mallon et al. (2010) note "The main difference between AAT and other adjunctive therapies is that the central 'tools' in this intervention are living, breathing, interacting creatures" (p. 135). This is an important difference: When animals are introduced into a health or mental health setting, unique organizational issues must be considered.

For the purpose of this chapter, animal-assisted interventions are defined as "any therapeutic intervention that intentionally includes or incorporates animals as part of the therapeutic process or milieu" (Kruger and Serpell, 2006, p. 36).

Initial Research

Friedmann et al. (1980) conducted a longitudinal study of patients with coronary heart disease and their exposure to pets. Their research indicated that one year after discharge, pet owners were more likely to be alive than those without pets. People who owned a pet had one-third the mortality rate of those who did not own a pet. Patronek and Glickman (1993) suggest that since coronary heart disease is a stress-related disease, the protective effect of pet ownership is due to its effect on psychological risk factors. They concluded that by reducing stress and improving mental health, companion animals may make their owners more likely to survive. The study conducted by Friedmann et al. was one of the first to indicate the positive effects of companion animals and sparked a new area of research.

In 1994, a group of Australian researchers conducted the first national study that investigated the relationship between pet ownership and human health (Headey, 1999). The results of their research indicated that dog and cat owners had better mental and physical health than those who did not own a dog or cat. Pet owners made fewer visits to the doctor and were less likely to be on medication for heart problems or for difficulties with sleeping. Headey (1999) concluded that owning a dog or cat probably reduces the national health expenditure.

Marguerite O'Haire (2010), citing Beck's and Glickmans's (1987) work in response to the growing body of data that indicates the positive effects of companion animals on human health, declared that "No future study of human health should be considered comprehensive if the animals with whom people share their lives are not included" (p. 227).

There have been many positive anecdotal examples regarding the use of animals, but it is important to note that there is limited empirical support and research that validates the effectiveness of animal-assisted interventions. One of the greatest challenges facing supporters of animal-assisted interventions is the fact that many of the professionals who apply these strategies do not see the necessity for conducting outcome research. This lack of documentation leaves a great void in demonstrating the efficacy of this kind of therapeutic intervention. Katcher (2000) and Serpell (1983) note that just because an interaction with an animal is enjoyable does not imply the procedure is therapeutic. So, the need for more empirical research is apparent in the field.

With the limited body of research that indicates the positive benefits of pet ownership, researchers began to wonder about the underlying mechanisms of these human-animal interactions. Many theories have been offered, but to date there is no unified empirically supported theoretical framework to explain how companion animals benefit the mental and physical health of human beings. Two theories that are most commonly cited are the biophilia hypothesis and the social support hypothesis.

Biophilia Hypothesis

The biophilia hypothesis contends that "humans have an innate propensity to attend to and be attracted by other animals and living things" (Wilson, 1984, cited in O'Haire, 2010, p. 227). O'Haire (2010) observes that it would have been evolutionarily beneficial for humans to pay close attention to animal behaviour as environmental indicators of safety or danger. Gunter (1999) notes that having a pet in the home may be a link to our human evolutionary history, one which enhances our psychological health.

Friedmann (1995) discovered in his research that when humans looked at animals, it decreased their anxiety. Friedman et al. (1983) studied the effect of a friendly dog on children's blood pressure while resting and during the mildly stressful activity of reading out loud. Researchers measured the children's blood pressure, both resting and reading, when the dog was present and when it was not, and they discovered that the children's blood pressure was lower when they were resting and reading while the dog was present. The researchers concluded that in mildly stressful situations the presence of a dog can reduce anxiety.

Katcher et al. (1983) did a research study measuring participants' blood pressure as a measure of their level of stress while participating in three activities: watching fish in a fish tank, watching a blank wall, and during the mildly stressful task of reading aloud. They discovered that watching the fish lowered the participant's blood pressure and produced a state of relaxation. In the participants with hypertension, the activity of watching fish in a tank lowered their blood pressure into the normal range. The authors concluded that watching animals, whether or not they were familiar or bonded to them, can reduce anxiety and tension.

Wilson (1991) summarized the physiological effects:

> Thus, it has been hypothesized that pets can decrease anxiety and sympathetic nervous system arousal by providing a pleasant external focus for attention promoting feelings of safety and providing a source of contact comfort. They can decrease loneliness and depression by providing companionship, promoting an interesting and varied lifestyle, and providing an impetus for nurturing. Pets, therefore, have the potential to moderate the development of stress-related diseases such as coronary heart disease and hypertension. The range of benefits that owners might derive from their pets may not pertain only to pet owners; one could speculate that anyone, not just pet owners could benefit from the presence of friendly animals. (cited in Odendaal, 2000, p. 275)

Research has also been done that indicates the presence of an animal can change a person's perception of the scene. Murray (1943) did the original research where participants were shown pictures of people in "provocative, yet ambiguous scenes" and were asked to describe the scenes. Lockwood (1983) and Friedmann and Lockwood (1991) expanded on this research using the Animal Thematic Apperception Test (ATAT), which uses two sets of pictures that are identical, except that one set of pictures has an animal present. Participants were asked to describe the scenes. People who were present in scenes with an animal were consistently described as "friendlier, happier and less threatening than the same people in scenes without animals" (cited in O'Haire, 2010, p. 228). The authors concluded that a person's perception of a situation influences their stress response, and if they are able to respond more calmly, this may enhance their psychological wellbeing.

Social Support Hypothesis

The social support hypothesis asserts that human-animal interactions have significant benefits. The lack of social support is commonly seen as a risk factor for both physical and mental health issues. Supporters of this hypothesis contend that companion animals are social supports unto themselves but also facilitate social interactions with others.

As social supports, companion animals reduce loneliness and often contribute to a more general sense of well-being for their owners. Pet owners talk about having a "reason to get up in the morning," "someone to come home to" or "somebody that helps structure my day." Friedman et al. (1980) and Kruger et al. (2004) noted that pet owners reported their companion animals provided social support because they were constantly available, were non-judgmentally supportive, and provided unconditional love. The result of these social interactions is the strong attachment people form to their pets.

Pet companions provide a source of pleasure, connection to the outside world and, for some people, the promise of hope and a reason to live (Fine, 2006). Norris et al. (1999) invited seniors to complete questionnaires relating to their life satisfaction and perceived health. Pet owners reported higher life satisfaction and perceived health than those who did not own a pet. The researchers cited the non-evaluative support provided by a pet as contributing to the more positive perception of their life and health by the owners. The researchers, however, did note that one of the limitations of this study was that people who were healthier and had a more positive outlook on life might also be more likely to own a pet and concluded that further research needed to be completed in order to determine if there was any causal direction in the relationship.

Gunter (1999) in his research proposed that companion animals were facilitators of social support between human beings, and that they acted as a kind of "social lubricant" that encouraged social interaction with others and often stimulated conversation. Eddy et al. (1988) conducted a study in which a participant in a wheelchair was sitting in a heavily trafficked pedestrian location. The person was either alone or with a service dog. An observer sat nearby to make note of the number and types of interactions between the person in the wheelchair and strangers. The results indicated that when a dog was present there were many more positive social approaches, including smiles and conversation, than when the dog was not present. Eddy et al. concluded that the presence of an animal could be helpful in overcoming social isolation because of the socializing effects of an animal. This author can attest to this phenomenon as well. Upon becoming the owner of a puppy and beginning to go for walks in the neighbourhood, many more neighbours said hello, and stopped me to initiate conversation than when I walked in the neighbourhood without the puppy.

Context of the Research

Wilfrid Laurier (WLU) is a medium-sized university that advertises itself as, "a multi-campus university with strengths in many academic areas and a focus on 'inspiring lives of leadership and purpose' in order to provide a great student experience" (www.wlu.ca). Like many universities in Canada, WLU has increasingly become concerned about the mental health needs of its students and how these impact their academic achievement. This concern has become particularly important after a series of suicides and incidences of self-harm on a number of Canadian university campuses. Greater awareness of mental health issues on campus is also the result of increasing de-stigmatization and more open discussion in the greater Canadian context. One in five Canadians will suffer from mental illness at some point in their life and according to Statistics Canada, youth aged 15 to 24, which includes most undergraduate students, are the most likely group to suffer the effects of mental illnesses, substance dependencies, and suicide. Data that was collected by WLU Counselling Services note that of the 802 students who accessed counselling services in the 2013–2014 academic year, 313 (39%) named anxiety as the primary reason for accessing the service, and 255 (32%) students named anxiety as the secondary reason for attending counselling. This means that a total of 568 (71%) of all students accessing counselling services named anxiety as their concern (Report by WLU Counselling Services, Client Goals Sept. 1, 2013 – Apr. 30, 2014). In a meeting with the counsellors at WLU Counselling Services, it was mentioned that many of the students also talked in their counselling sessions about how much they missed their pets.

Hanlon (2012) notes that "Anecdotal evidence suggests that today's students experience more stress due to higher levels of debt, fewer job prospects upon graduation, higher expectations for student success, and less preparation for independent living" (p. 2). Students' stress becomes more acute as they approach and are in the midst of writing final exams. Could interacting with dogs have a positive impact on students during a high stress time such as final exams? This question led to a research project that investigated the question: "What is the impact of interacting with puppies during final exams?" It received ethics approval from WLU REB #3998.

Methodology

This research project used a mixed methodology comprising of a survey and qualitative data collected regarding participants' experience of interacting with puppies. Participants responded to posters on campus and social media communications to come and interact with two puppies: Jager, an eight-month-old Soft Coated Wheaton Terrier, and Annie, a four-month-old Portuguese Water Dog. These dogs were chosen because both are hypoallergenic breeds, which would allow more students to interact with them with less concern about allergies.

Upon arrival, students were given an information sheet describing the study and an informed consent form to sign. Then they were asked to note their feeling(s) upon arrival and rank it/them on a Likert Scale, between one and five, where one was very low intensity and five was extremely high intensity. They were then free to interact with the dogs in whatever manner suited them. Some participants simply sat and watched the puppies play or interact with other participants. Some participants played with the toys available for the dogs, petted them, or gave them dog treats for completing a command (sit, down, shake a paw, high five/ten, etc.). Participants could stay as long as they wanted, with the majority of the participants staying between 10 minutes and one hour. Upon leaving, participants were asked to note their feeling(s) and rank it/them on the same Likert Scale. If the participants noted a change they were asked to describe what they thought contributed to the change. Finally, they were asked some basic demographic information, such as gender, age, faculty, and year of program.

Findings

In the study, 74 students (n=74) participated in the project: 60 females (81%) and 14 males (19%). This stands in contrast to the gender breakdown at WLU, where females comprise 54% of the university population and males 46%. Many more females than males chose to participate in the research project, and therefore, the percentages were not proportionate to the gender balance of the wider university population.

At the time of this study, 46 of the students (62%) had a pet; 52 of the students (70%) had a pet growing up; and 12 of the students (16%) did not grow up with a pet, nor did they currently own one. Since students self-selected to participate in the event, it would follow that

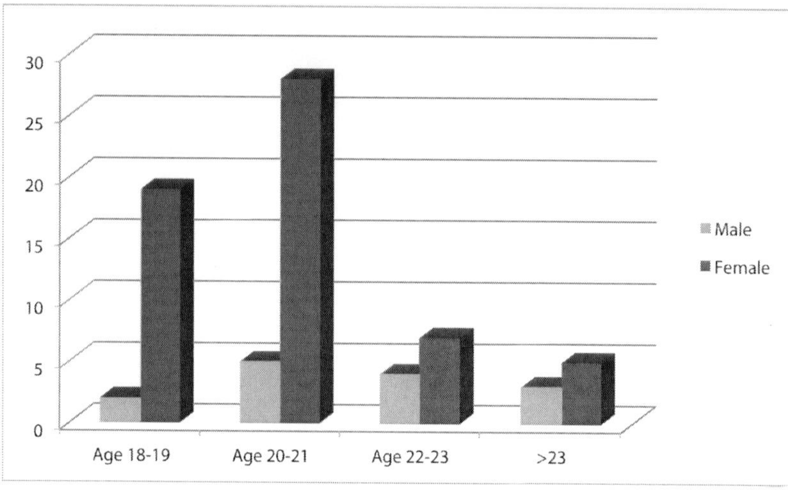

Figure 1. Participants by Age and Gender

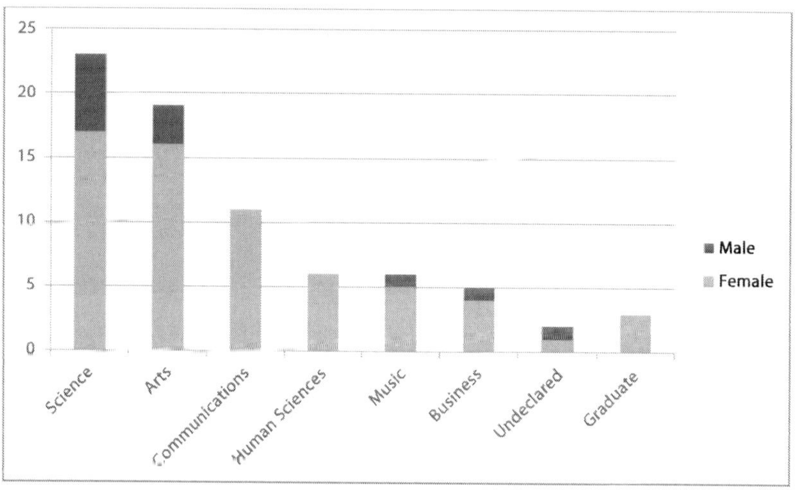

Figure 2. Participants by Faculty

students who had previous or current experiences with a pet would be more likely to participate.

Upon arrival, students were asked to observe what they were feeling and rate it on a Likert scale, where one indicated low intensity and five indicated extremely high intensity. They noted the following, with some students noting more than one feeling:

As students left, they were again asked to note what they were feeling and to rate their feelings on a Likert scale, from one to five,

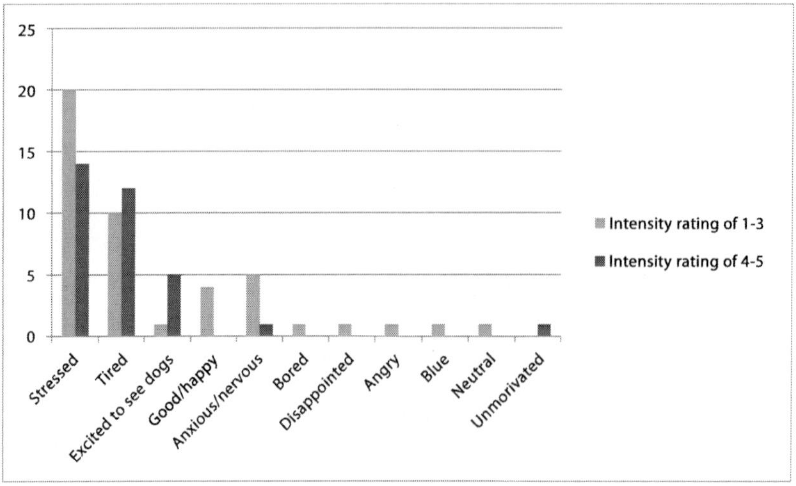

Figure 3. Participants' Feelings upon Arrival

where one indicated low intensity and five was extremely high intensity. Some students noted more than one feeling.

It was a surprise to the researcher that, on leaving, only nine of the students rated the feeling they noted upon arrival. In eight of these rankings, the feeling (stress) decreased in intensity, while one ranking remained the same. All the other feelings the students recorded when leaving were different from the feelings noted when they entered, which consequently did not allow the researcher to calculate

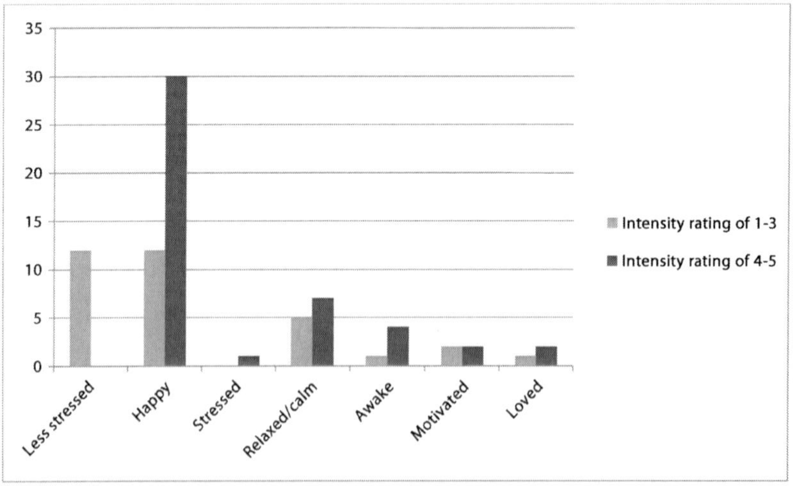

Figure 4. Participant's Feelings Upon Departure

the degree of change that was experienced by the participants. The six students (all female) who arrived excited to see the dogs reported feeling happy when they left. All six students had either owned a dog growing up or currently owned one. Only one student who arrived stressed (intensity of 4) left feeling the same level of stress (intensity of 4). The remaining 73 of the 74 students experienced positive change in their feelings.

Five themes emerged from the qualitative data: energy of the dogs; distraction from current stresses; touch or contact with the dogs; unconditional love; and a reminder of home or pet.

1. Energy of the Dogs

Many of the participants (34, or 46%) noted the puppies' energy, saying things like, "the dog's energy and happiness rubs off," "their energy is infectious," "dogs make me happy because they are playful," and "the joy from the dogs is contagious." Participants experienced the puppies' energy as affecting them, whether or not they physically interacted with them. Simply watching the dogs also "rubbed off" on the participants.

2. Distraction from Current Stress

Eighteen participants (24%) mentioned that the puppies served as a distraction from their current stress. They noted that "the puppies help you get your mind off things you may be worrying about when playing with them," "watching the dogs takes my mind off stressful exams ahead," "playing with the dogs gave me time to think of something else," "seeing the puppies just being carefree made my problems just seem to disappear," "playful puppies help me unwind or forget my worries," and "it helped me remember that school is not the end of the world."

3. Touch or Contact with the Dogs

A number of participants (12, or 16%) noted that touching the dogs made a difference to them. They observed the following: "The feel of the dog's fur and their cuteness makes me think nothing can go wrong," "connection with another creature helps," "being able to pet the dogs helped me," "playing with the puppies took my mind off my work and exams and helped me to relax and keep calm," and "being able to play and cuddle with the dogs helped me." It is interesting to note that while the sample size of male participants was small, all of them noted in their comments that physically interacting with the puppies helped them feel better. While this was true for some of the female participants, it was not mentioned by all.

4. Unconditional Love and Acceptance

Eight participants (11%) mentioned that it was the unconditional love and acceptance by the puppies that contributed to the change they experienced. They stated that "unconditional love was shown to me regardless of my grades," "puppies always love you," "puppy loving really relaxed me" and "they gave me unconditional love and attention even if I failed my exam." All participants who made comments regarding unconditional love and acceptance were female. None of the male participants mentioned anything regarding unconditional love and acceptance as being a contributor to the change they experienced.

5. Reminder of Home or Pet

Some of the participants (6, or 8%) commented on how interacting with the puppies reminded them of home or a pet they had when growing up. They stated that "the cuteness of the dogs reminds me of my dogs," "the animals remind me of home," "I love dogs, I really miss my dog, and it is nice to see them having fun and playing and interacting with me," and "they remind me of my puppy at home." While it might be expected that interacting with the puppies might make the students homesick, it rather seemed to be comforting in some way and helped relieve the current stresses. It may be a result of being familiar with dogs and previously having positive experiences with them that supported them to leave feeling more positive than when they arrived.

Discussion

The practice of using tame animals in therapy began in the eighteenth century when animals were introduced into mental institutions as a way to help socialize patients with mental disorders (Serpell, 2006). As noted earlier, it has only been very recently that researchers have begun to create standardized terms for their activities. The umbrella term "animal-assisted intervention" is defined by Kruger and Serpell (2006) as "any intervention that intentionally incorporates animals as part of a therapeutic or ameliorative process or milieu" (p. 36). The Delta Society, which is one of the largest organizations in the United States responsible for certification of therapy animals, has published the following definitions of animal-assisted therapy and animal-assisted activity:

> Animal Assisted Therapy (AAT): AAT is a goal-directed intervention in which an animal that meets specific criteria is an integral part of the treatment process. AAT is directed and/or delivered by a health/human service professional with specialized expertise, and within the scope of practice of his/her profession. Key features include specified goals and objectives for each individual and measured progress.
> Animal Assisted Activity (AAA): AAA provides opportunities for motivational, educational, recreational, and/or therapeutic benefits to enhance quality of life. AAAs are delivered in a variety of environments by specially trained professionals, paraprofessionals, and/or volunteers, in association with animals that meet specific criteria. Key features include absence of specific treatment goals; volunteers and treatment providers are not required to take detailed notes; visit content is spontaneous. (as cited in Kruger and Serpell, 2006, p. 34)

Given the drop-in nature of the research project and the spontaneous experience of the participants, the current research would be an animal-assisted activity rather than animal-assisted therapy. There were no specific goals developed for each of the participants, and the individuals involved in facilitating the event ranged from a trained professional to volunteers. The majority of students either had a pet at the time of the study or grew up with a pet, which predisposed their likelihood of engaging the opportunity to interact with the puppies. Also, a number of the participants who had not grown up with a pet nor currently owned one mentioned that they liked animals, with some of them noting they would have liked to have had a pet growing up and are looking forward to owning a pet, once their living conditions are more conducive to pet ownership.

The change in feelings that participants spoke about most often was related to the energy of the puppies. This would be in line with the biophylia hypothesis, which states that observing an animal has a positive effect on the individual's response to stress. The researcher observed that initially, when students entered the room, they were quiet and somewhat tentative about the experience. Once they were handed some treats to offer the dogs and encouraged to interact with them, the mood in the room significantly changed. Students began to laugh at the puppies' behaviour, marvelled at their ability to "high five/ten," and enjoyed playing with them. The same would occur for those students who chose to simply watch and not physically interact with the dogs. This would happen over and over again as new students kept arriving. The energy of the puppies seemed to be affecting the energy of the students.

The final exam period on a university campus is a stressful time, with students finishing papers/projects and studying for and writing exams. It is to be expected that students' stress levels would be higher; they would be more tired from staying up late, and they might feel overwhelmed or discouraged, depending on how they felt their exams were going. As noted in the biophylia hypothesis, it seemed to be of some benefit to the students to be able to focus on an external object (the dogs). They reported finding the distraction and break from thinking about their exams as helpful. Also, a few students who entered feeling tired and unmotivated noted that when they left, they were feeling more focused and motivated to go home and either study or finish some outstanding assignments. Serpell (1996) notes that it has been known since the 1950s that any stimulus which is attractive or which concentrates the attention has a calming effect on the body.

For many students, having physical contact with the dogs seemed to play an important role in the change they reported (Penkowa, 2015). This was particularly true for the male participants, an observation which leads this researcher to wonder whether there is a gender difference in what males and females might need from the experience. For these participants, it was not enough to simply watch the puppies play, but rather it was important for them to be able to pet or hug the dogs, play with them, and physically interact with the puppies. They noted the importance of the physical connection with another creature, which is often absent on a university campus or in university housing.

A number of the females noted the importance of experiencing the unconditional love and acceptance from the dogs. Certainly, much has been written regarding the human tendency to anthropomorphize our relationships with pets, projecting human feelings, motives and qualities onto animals. Serpell (1996) argues that most pet owners believe their pets "love" or "admire" them. He suggests that without this belief the relationships that most people have with their pets would be essentially meaningless. Therefore, it was interesting that a number of participants (all female) made reference to such humanistic qualities as unconditional acceptance and love which they "experienced" from the dogs. This was very important to them in contributing to their experience of change in their feelings. As noted above, it is interesting that this was not mentioned at all by any of the male participants.

The final theme noted by the students related to the dogs reminding the participants of home or a pet. There seemed to be some comfort garnered by the familiarity of interacting with the dogs. All the participants who mentioned this had grown up with a dog, and some of them still had a dog at their parent's home. The researcher initially thought this might increase the student's distress or homesickness but was surprised to note that it actually had the opposite effect. Perhaps what also contributed to the positive change was that these participants had experience in using the relationship with their pet to modify or ameliorate difficult feelings, so interacting with the two puppies provided a means to access a familiar resource.

It is curious that there were so many more females than males (81% to 19%) who chose to participate in the opportunity. One possible contributor to this imbalance might be the way the event was advertised. The poster had a picture of the two puppies together with the questions: "Feeling stressed? Need a little love?" Perhaps young men on a university campus do not respond to such an invitation as possible relief for their stress.

It will be important in future research to include another question that asks participants to both rate the feeling they arrived with and also that same feeling upon departure. Another question could also be added, asking what they are feeling at departure. This way the degree of change could be calculated. Also, it would be interesting to follow up with the participants the following day to ascertain how long the change lasted. Was it transitory, or did the change last for a longer period of time? Is there any correlation between how much time the participants spend with the dogs and the degree of change, and how long the change lasts? In preparation for a future event, the researcher could speak with some males attending university and ask about what type of advertising might be more attractive to incite them to participate in the project. This might encourage a higher percentage of male participants. How important to the outcome was the fact that it was two puppies that interacted with the students? Would adult dogs elicit the same response? These questions will be important to consider in future research.

Conclusion

While the sample size for this project was relatively small (n=74), 73 participants indicated positive change in how they were feeling as a result of interacting with the puppies. Since final exams tend to be a highly stressful time for students, offering an opportunity to interact with an animal could be an important resource for students to manage their stress. This resource is of particular importance to students who have had previous positive experiences with animals. With the increase in students presenting with mental health issues and greater attention being paid to these issues on university campuses, providing animal-assisted activities may be a low-cost resource to support students and promote a greater sense of well-being.

References

American Veterinary Medical Association. Wellness guidelines for animals in animal-assisted activity, animal-assisted therapy and resident animal programs; 2011. Retrieved from http://www.avma.org/KB/Policies/Pages/Wellness-Guidelines-for-Animals-in-Animal-Assisted-Activity-Animal-Assisted-Therapy-And-Resident-Animal-Programs.aspx

Archer, J. (1997). Why do people love their pets? *Evolutionary Human Behavior, 18,* 237-259. CBC News. Retrieved from http://www.cbc.ca/news/canada/doggy-spas-and-gourmet-food-drive-canadian-pet-costs-higher-than-ever-1.2577156

Consumer Corner: Canadian Pet Market Outlook, 2014. Retrieved from http://www1.agric.gov.ab.ca/$department/deptdocs.nsf/all/sis14914

Clutton-Brock, J. (1995). Origins of the dog: domestication and early history. In J. Serpell (Ed.), *The domestic dog: Its evolution, behaviour and interactions with people* (pp. 7-20). Cambridge, UK: University of Cambridge.

Davis, S. J. and Valla, F. R. (1978). Evidence for the domestication of the dog 12,000 years ago in the Natufian of Israel. *Nature, 276,* 608–610.

Eddy, J., Hart, L. A. and Boltz, R. P. (1988). The effects of service dogs on social acknowledgments of people in wheelchairs. *The Journal of Psychology, 122*(1), 39–45.

Fine, A. H. (2006). Incorporating animal-assisted therapy into psychotherapy: Guidelines and suggestions for therapists. In Fine, A. H. (Ed.), *Handbook on animal-assisted therapy: theoretical foundations and guidelines for practice* (pp. 167–206). San Diego, CA: Academic Press.

Friedmann, E. (1995). The role of pets in enhancing human well-being: physiological effects. In Robinson, I. (Ed.), *The Waltham book of human-animal interaction: Benefits andresponsibilities of pet ownership.* Oxford, UK: Pergamon Press.

Friedmann, E. and Lockwood, R. (1991). Validation and use of the animal thematic apperception test (ATAT). *Anthrozoos, 4*, 174–183.

Friedmann, E., Katcher, A. H., Lynch, J. J. and Thomas, S. A. (1980). Animal companions and one-year survival of patients after discharge from a coronary care unit. *Public Health Report, 95*(4), 307–312.

Gullone, E. (2000). The biophilia hypothesis and life in the 2lst century: Increasing mental health or increasing pathology? *Journal of Happiness Studies, 1*, 293–321.

Gunter, B. (1999). *Pets and people: The psychology of pet ownership.* London, UK: Whurr Publishers Ltd.

Hanlon, C. State of Mind: Addressing Mental Health Issues on Campus. *University Manager*. Summer, 2012. Retrieved from http://www.aucc.ca/wp-content/uploads/2012/06/mental-health-state-of-mind-university-manager-article-summer-2012.pdf

Headey, B. (1999). Health benefits and health cost savings due to pets: preliminary estimates from an Australian national survey. *Social Indicators Research, 47*, 233–243.

Katcher, A. H. (2000). Animal assisted therapy and the study of human-animal relationships: discipline of bondage? Context or transitional object? In A. Fine (Ed.), *Handbook on animal-assisted therapy.* San Diego: Academic Press.

Katcher, A. H., Friedmann, E., Beck, A. M. and Lynch, J. J. (1983). Looking, talking and blood pressure: The physiological consequences of interactions with the living environment. In Katcher, H. H. and Beck, A. M. (Eds.), *New perspectives on our lives with companion Animals* (pp. 351–359). Philadelphia: University of Pennsylvania Press.

Kruger, K. A. and Serpell, J. A. (2006). Animal-assisted interventions in mental health: Definitions and theoretical foundations. In Fine, A. H. (Ed.), *Handbook on animal-assisted therapy: Theoretical foundations and guidelines for practice* (pp. 153-157). San Diego, CA: Academic Press.

Kruger, K. A., Trachtenberg, S. W. and Serpell, J. A. (2004). Can animals help humans heal? Animal-assisted interventions in adolescent mental health. Center for the Interaction of Animals and Society (CIAS) and University of Pennsylvania School of Veterinary Medicine, Philadelphia, PA.

Lockwood, R. (1983). The influence of animals on social perception. In Katcher, A. H., Beck, A. M. (Eds.), *New perspectives on our lives with companion animals* (pp. 64–72). Philadelphia: University of Pennsylvania Press.

Messent, P. R. and Serpell, J. A. (1981). An historical and biological view of the pet-owner bond. In Fogle, B. (Ed.), *Interrelations between people and pets* (pp. 5–22). Springfield, IL: Charles C. Thomas.

Mallon, G. P., Ross, S. B., Jr., Klee, S. and Ross, L. (2010). Designing and implementing animal-assisted therapy programs in health and mental health organizations. In A. Fine (Ed.), *Handbook on animal-assisted therapy: Theoretical foundations and guidelines for practice* (pp. 135–147). San Diego, CA: Academic Press.

Murray, H. A. (1943). *Thematic apperception test manual.* Cambridge, MA: Harvard University Press.

Norris, P. A., Shinew, K. J., Chick, G. and Beck, A. M. (1999). Retirement, life satisfaction, and leisure services: The pet connection. *Journal of Park and Recreation Administration* Vol. 17, No. 2, 65–83.

Odendaal, J. S. J. (2000). Animal-assisted therapy- magic or medicine? *Journal of Psychosomatic Research, 49*, 275–280.

O'Haire, M. (2010). Companion animals and human health: Benefits, challenges, and the road ahead. *Journal of Veterinary Behavior, 5*, 226–234.

Penkowa, M.D. (2015) Dogs and human health: The new science of dog therapy and therapy dogs. Bloomington, IN: Blaboo Press.

Patronek, G. J. and Glickman, L. T. (1993). Pet ownership protects against the risks and consequences of coronary heart disease. *Medical Hypotheses, 40*, 245–249.

Serpell, J. A. (2006). Animal-assisted interventions in historical perspective. In Fine, A. H. (Ed.), *Handbook on animal-assisted therapy: Theoretical foundations and guidelines for practice.* San Francisco: Academic Press.

Serpell, J. A. (1996). *In the company of animals.* (2nd ed). Cambridge: Cambridge University Press.

Serpell, J. A. (1983). Pet psychotherapy. *People-Animal-Environment* 7-8.

Wilson, C. C. (1991). The pet as an anxiolytic intervention. *Journal of Nervous and Mental Disease, 179*(8), 482–289.

Chapter VII

Inter-Partner Touch in Couple Counselling: Theory and Emerging Practice

Cassandra Petrella and Martin Rovers[1]

Appropriate touch between a health practitioner and patient/client has long been understood to promote safety, comfort, and healing. Appropriate touch between a therapist and client in individual counselling is also well documented as contributing to client growth and trust in therapy. However, in the field of couple counselling, the benefits related to use of inter-partner touch has received little attention. Some consideration has been given to at-home touch outside of session, in particular when one partner is a survivor of abuse (Maltz, 1995). However, touch between partners in session has been largely ignored. What is inter-partner touch in counselling, and what needs to be researched about these interventions in couple counselling intervention? This review serves as a call towards future research in this forgotten area of human development. An emerging model for integrating inter-partner touch into existing clinical models is proposed.

What is Touch?

Defining Touch

Touch is a bodily sense. It is also a proximal sense: "that is, we feel those things that either are close to us or actually contact us" (Cholewiak and Collins, 1991, p. 23). We feel the warm touch and tactile pressure as a partner reaches out to hold our hand or hug us. Our skin, which conveys the sensations of touch, is our largest sense receptor organ (Montagu, 1971). Our entire body is capable of receiving touch. Our skin receives information and messages about our environment, our connection to others and, ultimately, our felt sense of being loved and secure. Montagu believed touch to be the strongest sense.

"Inter-partner touch" will be the term used for the touch which occurs between partners within a couple counselling session or elsewhere. Touch between partners can be eye-to-eye contact, hand-to-hand touching, embracing, or hugging. The touch we refer to is mostly non-sexual. We include close eye-to-eye contact as touch in our definition, based on the biological belief that body energies join when the energies meet in close proximity (Gerber, 1996). As this is a new area of therapy and research, inter-partner touch has been created for the purpose of this paper and future research. So why is inter-partner touch so crucial for couple love and security?

Psychobiological Function of Touch

The components of the brain that are of interest in understanding touch are the primary and secondary somatosensory cortexes in the cerebrum. This section of the brain processes touch signals from various parts of the body; it can be activated by mere observation of touch and it integrates sensory information in terms of temperature and pressure to construct an understanding of the object that is being encountered or touched (OECD, 2007). Deep within the cerebrum is the limbic system, which is often referred to as the emotional brain.

The limbic system contains two key brain regions necessary for touch: the hypothalamus and the amygdala (OECD, 2007). The hypothalamus is the source of feel-good chemicals and allows for the interplay of emotion and reason. The amygdala is our alarm system responsible for responding to non-verbal cues of threat, anger, avoidance, defensiveness, and fear, and it provides the integration of emotional perception and experience (Solomon and Tatkin, 2007). Both make it possible for humans to experience both healing and wounding touch.

Brain plasticity makes it possible for touch to create shifts in a couple's interactions over the long term. "Changes in neural pathways can occur as a result of behaviour, environment and neural processes; based upon positive experiences with touch, neural pathways can change" (OECD, 2007, p. 165). A person's experiences can actually change both the brain's physical structure and the brain's functional organization. Touch can create increases in feelings of safety and security, which leads to an increase in connections between the left and right hemispheres of the brain (Solomon and Tatkin, 2007). It follows, then, that the use of touch in couple counselling can be used proactively as an intervention to remove fear, repair attachment wounds of childhood, and increase feelings of safety. Touch can lead the person to think, act, and feel secure in the couple relationship.

What is the Importance of Touch for Psychological Well-Being?

Touch Promotes Human Development

All humans and animals have an innate and vital need for touching from infancy.

> Touch is the earliest sensory system to become functional in all species [...]. Perhaps next to the brain, the skin is the most important of all our organ systems. The sense most closely associated with the skin, the sense of touch, is the earliest to develop in the human embryo. (Montagu, 1971, pp. 1–2)

A mother's grooming, licking, petting, and holding are all seen as crucial to infant development in many species (Montagu, 1971). Harlow (1958), in his experiments with rhesus monkeys, demonstrated how they valued the touch and feel of their cloth mother more than food. Monkeys developed brain damage, became violent, and socially impaired as adults when they were deprived of "contact comfort."

A child needs constant tactile proximity with a caregiver in order to assure survival (Mikulincer and Shaver, 2010). Similar to Harlow's findings, Spitz (1945) found that the infants deprived of physical contact did not develop normally. He noticed that when orphans were provided with the minimal basic needs of feeding and diaper changing but without ongoing comfort and human contact, by two months, the infants' development became delayed. Later, Bowlby would name this deprivation in the three stages of protest, despair, and detachment (Bowlby, 1969). Modern researchers, such as Dadds et al. (2012), consistently demonstrate that the absence of eye-to-eye contact in childhood impairs the facilitation of growth in discipline, affection, and empathy. Such repeated findings imply that, as well as and beyond meeting a person's basic needs, human beings need touch in order to survive.

Field et al. (1987) demonstrated the importance of touch for preterm infants in massage therapy studies. Infants who were in the neonatal intensive care unit were massaged three times a day for 10 days at a period of 15 minutes each time. Compared to a control group of pre-term infants who were not massaged, the researchers found that the massaged infants gained 47% more weight and were released from the hospital on average six days earlier. In follow-up studies with the infants at one year of age, the infants who were massaged were still at a weight advantage and had higher ratings on tests of mental state and motor skills. In another study by Field (2002), mothers reinforced infants' eye contact though either facial expressions, voice and/or touch. The infants who received touch reinforcements were found to smile more, vocalize better, and cry less, which Field describes as characteristics of a developmentally healthy and happy baby.

Touch Provides Attachment Security

Touching is said to be the language of attachment, security, and comfort (Bowlby, 1969; Montagu, 1978; Mikulincer and Shaver, 2010). From contact with a mother, father, or caregiver, the child develops his or her attachment security. When the infant is in need of care, they seek proximity to their caregiver. When the infant is hungry, the child will cry and reach his or her hands out to the caregiver (Ainsworth, 1989). When proximity attempts are met consistently by the caregiver, the child will develop a sense of a secure base in the relationship. With a secure and trusting relationship base, the child can meet developmental milestones by feeling comfortable in exploring his or her environment and interacting with new individuals. The child understands that he or she can approach the unknown and return to a secure base for protection or support if needed. Experiencing consistent caregiver availability leads to future positive social beliefs such as feeling comfortable with intimacy, interdependence, trust, and affection (Mikulincer and Shaver, 2010).

If the caregiver is frequently unavailable or inconsistent when proximity is sought, attachment wounds will develop within the child. When attempts for proximity seeking to a primary attachment figure are not met and pain is experienced, secondary coping strategies need to be elicited in order to self-regulate distress (Ainsworth,

1989). An infant will either learn to deactivate or hyperactivate behaviour and emotions in an attempt to self-soothe. The decision to choose either strategy is based on the infant's interpretation of the caregiver's lack of attention (Mikulincer and Shaver, 2010).

Ainsworth outlined two kinds of attachment wounds that occur for the infant: avoidant and preoccupied. An avoidant or deactivating strategy is chosen based on failure of attachment behaviours to achieve proximity and on being punished for seeking proximity. The caregiver disapproves of closeness and contact. With such an experience, the caregiver is seen as dangerous, and the infant is forced to choose deactivating coping strategies in order to protect from further pain. The infant often becomes self-reliant and auto-regulating.

Alternatively, a hyperactivating or preoccupied attachment strategy is chosen when the co-regulation of distress is attempted and the caregiver fails. When a parent acts inconsistently or inappropriately, an infant feels that he or she is unable to act autonomously and must work harder to gain proximity. Distance from the caregiver is seen as dangerous, and the infant becomes afraid of what will happen if he or she tries to cope and regulate distress. The infant learns to seek proximity more aggressively and urgently.

The child who does not develop secure attachment will develop an attachment wound and will rely on the secondary attachment strategies in order to gain proximity or security with others. Later, in couple relationships, these secondary strategies can lead to negative interaction experiences with a partner. They are cause for relationship distress. For example, an adult partner with avoidant tendencies is not comfortable with touch and proximity due to his or her childhood wounds where touch led to rejection. Closeness for the avoidant person brings up feelings of past disappointment, shame and emotional pain (Solomon and Tatkin, 2011). The person will be uncomfortable with both giving and receiving intimacy. Studies have shown the links between an avoidant attachment strategy and distancing in relationships (Feeney, 1998; Shapiro and Levendosky, 1999).

On the other hand, the preoccupied person will seek closeness to such an extreme that their partner becomes uncomfortable with it. The preoccupied individual often misinterprets a normal degree of closeness as a partner being disengaged and disinterested in them. Preoccupied individuals become insecure and distressed when their bids for more closeness are not met consistently (Mikulincer and Shaver, 2010). The individual misinterprets a partner's physical behaviours as personally rejecting and starts to fear the end of the relationship. Such behaviour would be particularly difficult for an avoidant partner who is uncomfortable with even normal levels of closeness. The pull of such a relationship would be called a "pursuit-withdraw negative interactional pattern" (Johnson, 2008). Couples often seek counselling when they are experiencing this common couple interactional pattern, as it leads to much distress and dissatisfaction in relationships: childhood attachment needs are unmet, partners' responses are misunderstood, and neither partner knows how to repair the childhood wounds.

For individuals who are secure, avoidant, or preoccupied, their views of touch in relationships all differ. In a study on attachment styles and closeness in relationships, Feeney and Noller (1991) asked partners to describe their relationship. Content analyses revealed that secure people find it important to have a balance between closeness and independence, while avoidant people limit closeness and preoccupied people note closeness as more important than independence. Such differing and maladaptive views of touch can often lead to feeling unappreciated and dissatisfied in a relationship.

We each have an attachment wound or two, usually born in the family of origin in our relationship with our significant caregivers (Rovers, 2005). These childhood attachment wounds, such as avoidance or preoccupation, will often be a straightforward continuation into adult couple relationships (Bowlby, 1969). Healing childhood attachment wounds within the couple relationship becomes a core direction for couple counselling, and using inter-partner touch within the counselling session can be an integral part of partner healing.

Touch for Comfort and Well-Being

Biologically, touch regulates our internal homeostasis and well-being. Touch creates feel-good chemicals in the body. Touch stimulates the production of oxytocin and improves social bonding (OECD, 2007). Oxytocin is produced in the mother and in the infant when a baby is nursing. Oxytocin is often referred to as the "love hormone," as it calms and counters stress and promotes our desire to be touched. Dopamine is also produced through touch and serves as a teaching signal to avoid punishment and seek rewards, allowing a person to distinguish between safe and unsafe touch (OECD). These feel-good chemicals reinforce our decision to seek touch based on our innate needs for psychological well-being.

In a study at the University of North Carolina (Grewen, Girdler, Amico and Light, 2005), researchers looked at perceived partner support by examining the relationship between warm partner contact and oxytocin, cortisol, norepinephrine, and blood pressure. They examined self-reported partner support before and had a warm inter-partner contact moment. They found greater self-reported partner support was related to higher oxytocin levels for both genders. For women, a perception of high partner support correlated with lower systolic blood pressure after warm contact but not before the contact. Also, higher oxytocin was related to lower systolic blood pressure levels and lower norepinephrine for individuals in the study. These findings support the claim that positive connection and touch in a relationship can lead to better health for an individual.

Touch in Times of Distress

Mikulincer and Shaver (2010) explain that helping a partner emotionally regulate in times of distress is key to healing attachment wounds and creating a secure adult attachment style. Distress can be reduced through interactive regulation; partners can assist each other in reducing stress, tension, and fear (Mikulincer and Shaver, 2010). Interactive regulation is first experienced in infancy. The still-face experiment (Tronick, 2003), wherein an infant's caregiver keeps an unexpressive face in front of the child, causes great distress to the infant. The child is used to active interaction and interactive regulation with the parent and becomes upset by the parent's non-interactive response. The same distress would be expected to occur in couples when one partner is not reciprocating touch.

Inter-partner touch has also been found to reduce distress in times of immediate threats from a partner. In a study by Coan, Schaefer and Davidson (2006), participants lying in an MRI brain scanner anticipating a painful blast of white noise showed heightened brain activity in regions associated with threat and stress. But participants whose romantic partner stroked their arm while they waited for the white noise did not show this reaction at all. The pain seemed less frightening when their partner was touching them. Partners seem to be noticeably soothed and relieved through inter-partner touch. Couple counsellors can recreate couple reduce inter-partner stress through the use of inter-partner touch.

Touch Leads to Spiritual Health, Experience, and Connection

Spiritual mentors and healers often used touch, like "the laying on of hands" or hand holding, as part of their spiritual exercise and prayer. A study by Engebretson and Wardell (2012) examined the link between energy therapies and spiritual experiences. Participants were asked questions to describe their experience related to past energy-based healing techniques. Of the respondents, 17% mentioned a spiritual experience or interpretation from an energy-based healing moment. Similar spiritual descriptions were given by some of the healers who were offering the touch. These findings imply that for couples, a similar spiritual connection may develop with prolonged intimate touch. Such a connection could lead to spiritual growth between the couple and for the individuals.

The couple relationship itself can lead to "self-fulfillment through commitment to and trust in a partner [which] is one of the primary ways that people develop their innate potentiality to become spiritually complete and satisfied with life" (Berg-Cross, 2001, p. 341). The couple relationship that is rich in touch and connection can nourish spirituality.

Touch in Couple Counselling

Prior to the emergence of psychoanalysis, techniques involving touch were involved in the majority of psychiatric treatments (Levitan and Johnson, 1986). When psychoanalysis developed, Freud initially used touch with his patients. However, there were theoretical underpinnings which lead to removing touch all together from the treatment process. Then, as the field of psychotherapy advanced, legal and ethical worries developed that centered on protecting clients and oneself from litigation. Touch started to be feared as an avenue for a lawsuit or sexual misinterpretation. Such fears were passed down to the current socialization of counsellors, and today touch is mainly avoided in psychotherapy.

Weber's (1990) words that "touch as an interactional modality has been neglected," appears to be accurate when reviewing the literature. No empirical studies could be found that discuss the use of touch techniques in a couple counselling session. There are, however, several types of couple therapies that do involve some aspect of touch.

Touch in Today's Couple Counselling

Touch exercises can be integrated in couple counselling for special issues, for example, when a partner has been a victim of past sexual abuse. The previously abused partner must relearn safe touch, which they can do with the help of a supportive partner (Maltz, 1995). An abused person is often unfamiliar with what normal and appropriate touch feels like. They must learn to trust and enjoy touch with their partner while overcoming their sexual trauma through the process of counselling. Some of the nonsexual touch exercises involve placing a hand on their partner's heart while they gaze into their eyes, or writing messages on each other's backs. Eventually, the partners will move into an intercourse phase once the couple has become comfortable with all the other levels of touching. Many of the therapeutic techniques used with sexual abuse survivors could be used with other couples who are relearning touch after a period of disconnection.

Sensate focus (Masters and Johnson, 1970) therapy is another type of special issues therapy. Sensate focus is typically used for couples experiencing sexual difficulties. Partners are instructed to use minimal verbal communication while exploring each other's bodies and are to be guided by what interests them. This therapy is used in particular when a partner is experiencing anxiety or difficulties in sexual performance. The objective is to diminish sexual fears through a gradual development of comfort and understanding of touch with no pressure to perform. By the time the couple is ready to enter the sexual sensate focus period, they should be comfortable and at ease with touching each other. By creating comfort with touch through gradual activities, the couple relationship connection is strengthened.

Emotion Focused Couples Therapy (EFT) (Johnson, 2008) often asks couples who are experiencing sexual difficulties to abstain from sex while increasing their non-sexual touch at home, in order to increase their emotional connection. For couples to develop a healthy and satisfying sex life, Johnson attests that safe and non-pleasure demanding touch must be present. According to Johnson, touching, caressing, stroking and the emotions there evoked "are the royal route to love" (p. 191).

Solomon and Tatkin's (2011) new psychobiological approach to couple therapy (PACT) often involves the use of inter-partner touch in a couple's counselling session in order to connect with a partner at both the psychological and biological level. They explain that "gentle contact of skin to skin touch has a powerful effect on people. It is a reminder of the most basic human contact, that of child and caretaker, and often brings up strong emotion" (p. 29). They believe interactive regulation is key to regulating the emotional distress that has brought the couple to therapy.

In their therapy, Solomon and Tatkin use techniques such as eye-to-eye contact, face painting, "Toward and Away," and "Direct Embracing" in therapy to develop comfort with proximity and touch between partners. The partners are both asked to spend time reflecting on the sense of touch that is experienced. The therapist also monitors the couple's reactions to the touch and listens to their experiences. Attachment wounds and fears related to touch that come up are discussed and processed in therapy in order to overcome any aversion or difficulties to touch that were developed from past experiences. Partners eventually become more attuned to each other's internal experience.

Additionally, Solomon and Tatkin have their couples directly embrace in therapy. As the couple hugs, the therapist notes their reaction to such intimate touch and sees clues of their attachment styles. For example, an avoidant person can be seen to look away, hug with only one arm, not being fully engaged, or comfortable with the close contact. An anxious person might tense up as if they are expecting rejection by their partner. The therapist would explore such observations with the partners in order to deepen their awareness of each other's nervous system fear reactions. Through such exploration, one can begin to name and heal past attachment wounds that are affecting the current relationship. Solomon and Tatkin explain that "psychobiological management of two connected nervous systems [is] the most efficient manner of mutual regulation" (p. 168). The therapists believe that the quickest and most effective way to alter one state is to affect the state of your partner through touch.

A Spirituality for Inter-Partner Touch

Spirituality has three main components (Rovers and Kocum, 2010). For many, spirituality is about attunement with God or some Higher Being, a dimension of human experience that includes awareness of and a yearning for connection with this Being. For others, spirituality is about meaning of life: an experience where people try to make sense of the issues of life and find purpose in these life experiences, including love, and the loss of love. But for couples, spirituality is also about being in a relationship with others, the dimension of connection with others, sometimes a religious group, but more so connecting with each other. Most religions state that partners have the power to marry each other: a sacrament they bestow on their partner. The spiritual graces in marriage are to enter into vulnerability with each other; to know each other in deeper ways; to recreate life together; and to perfect the practice of forgiveness.

Touch is crucial to this spiritual grace for couples. Touch is a central and affirming part of most spiritual expressions: the expression of love; the laying on of hands in prayer; touching as a sign of healing; the embrace of forgiveness. Inter-partner touch has much to do with the meeting, loving, sexual, recreating, and caring for each other and children. Touch also plays a central part in the forgiving and healing dance of being in relationship. When partners deeply connect through touch in vulnerable moments of therapy, childhood wounds can be deeply explored and healed within the safety that touch provides. These moments have a spiritual essence in the beauty of the connection and healing that occurs through the outward expression of love and support between partners. That touch plays such an important role in couple relationships needs to be better expressed and further researched but, without doubt, touch is the sacramental, outward sign of the spirituality of love relationships. Whether it is through eye-to-eye contact, loving words, touch, or the lovers' embrace, touch is at the core of spirituality in love relationships.

An Emerging Model for Touch in Couple Counselling

The authors have been using inter-partner touch for some years now. We often administer the Inter-Partner Touch Scale (Rovers and Petrella, 2013) and the Dyadic Adjustment Scale (DAS) (Spanier, 2007) in the beginning sessions of therapy. The Inter-Partner Touch Scale is a measure developed to assess the frequency of touch in the relationship from each partner's perspective. The scale provides a sense of how often the couple is currently touching and is an indicator to the therapist about pacing and the need for psychoeducation on the benefits of touch. The DAS is administered as a measure of overall adjustment of each partner in the relationship. Both of these measures are re-administered around the sixth to eighth session to assess the couple's progress. Comparing the results of both assessments provides an indication of the effectiveness of the therapy course so far. In addition to these two measures, attachment styles and wounds are assessed via discussion in the individual sessions with the clients to provide the necessary background on what each partner brings to the relationship dysfunction.

Look, Touch, Talk, Hug

In our couple therapy work, we are using a new inter-partner touch therapy model, which combines the traditional components of inter-partner talking with new interventions of inter-partner touch. A case example will illustrate the new type of therapy we are using. We are seeing effectiveness in reducing both in-session distress and long-term de-escalation of conflict couples. Our couples also appear to be moving more quickly through the phases of developing secure attachment to each other. Worded another way, in the same way a caregiver held me when I was a baby, made eye contact, whispered loving words, and hugged me that created my secure attached self, so too the failure of this to happen created a wound or two in my insecure attached self. Our model of couple therapy creates a reenactment of secure attachment building through look, touch, talk and embrace.

As therapy begins and partners begin to uncover their primary emotions and internal experience of their relationship, it is important that partners stay engaged as their partner shares their personal experience of the relationship. As wounds are uncovered, the counsellor needs to create a new sense of safety with each partner. It takes time and safety for the wounds that were created in earlier childhood relationships to be named, claimed, and healed through the intimate experience of couple therapy. As partners begin to share their wounds with each other, we encourage the other partner to turn and look their partner gently in the eye and reassure them that they are here. Of course, this can take time, depending upon the nature of the childhood wound. Some partners' wounds are "look wounds," and so they are fearful to make much eye-to-eye contact, or fear to sustain such a look for long.

Susan is a 35-year-old woman, and, at best, she can make passing eye glances at her husband, Joe, whom she states she loves. A review of childhood experiences reveals that she was hospitalized at birth with a serious illness, and subsequently remained in hospital for several weeks each year. She remembers the bright lights and strangers' eyes looking at her, while feeling the fears and insecurities and pains of the experience. Susan has become wired to fear eye contact, as it means pain to her. When she could eventually name and claim this wound, Susan began to trust Joe's glance longer, and the therapy process in greater depth. After all, Joe is not one of those childhood caregivers who hurt her, but within Susan's emotional wiring, there are times when Joe's actions or words could replicate the same childhood pain and rejection.

As the work gets closer to the core attachment wounds, we encourage eye contact to be accompanied by other forms of touch, like a brief touch on one's partner's shoulder or hand holding. In one heated argument in which Susan was becoming flushed with anger, recounting an incident where Joe left for an evening meeting without saying goodbye to her, Joe is invited to touch Susan's hand, slowly, gently, lovingly, while looking in her eyes. Here Susan's permission is asked and Joe is invited to hold her hand while clarifying his explanation again that he thought she was napping and had no intention of hurting her. Over several sessions, Susan begins to trust the healing of look and touch.

The third component of the model is talk. The talk component is vital here for two reasons: first, it is necessary to name and claim the *wound word* that is at the bottom of fear, insecurity and mistrust; and second, it is equally crucial to name and claim the *healing word,* or soother and comforter, that can become the cornerstone of all one has ever wanted or needed to hear in terms of connection, comfort, and love. Susan names her wound word as "not heard." During the adult attachment interview assessment process early in therapy, Susan relates how she hates it when Joe does "not hear" her. It reminds her of the ways her parents, nurses, and other caregivers would ignore her cries to stop doing things, and just let her go home. Susan relates that Joe makes decisions without consultation, and explains decisions as a done deal, without talking to her. When the wound of "not heard" comes to visit Susan, she becomes angry and distant and she cannot even look at him.

Joe acknowledges that he is poor to listen, and that he has breached Susan's trust many times when he would just go somewhere without informing Susan, or without asking her opinion. He is used to being his own person right from childhood, and he rarely even asked his parents for permission. Since all have a wound or two, Joe's wound word is "my way," so when Susan's "not heard" dances with

Joe's "my way," the couple's dance of wounds is on and the negative spiral begins.

Healing words can change this negative dance, and help rewire both partners into a new dance of healing. Susan comes up with "hear me!!!" as her healing word, in the sense that she demands to be heard, and for Joe to consult with her. Joe comes up with "our way!!!" as his healing word in the sense of accepting his need to include others, especially Susan, and to begin the practice of asking for Susan's opinion. Becoming mindful of their old dance of wounds and turning to their healing words is difficult; and, oftentimes, healing words get left on the side of their couple road, leading to another fight. Healing takes much work, mindfulness, words, and especially touch. The couple is also asked what type of touch would feel soothing when a wound is triggered in the relationship. For example, one partner might say holding my hand while another partner might say rubbing my shoulder. The couple starts to learn what look, touch, or word feels most soothing to each person in times of emotional pain. This information is noted by the therapist and then specifically encouraged in session during particularly difficult therapy moments.

Soon, when the relationship feels safer, partners are asked to hug each other as a means to create bonding while still recognizing their wounds. As the partners gradually increase their awareness of the childhood wounds, support through various ways of touching is gradually increased. These stages of rewiring for love can occur in one session but most often take several sessions, depending on the couples' level of woundedness and depth of sharing. The intention here is gradual trust development, which is key to the development of secure attachment. The inter-partner touch encourages partners to know that it is okay to feel the way that they do and that they will be supported. Touch reassures them they will not be abandoned by their partner as they share these wounded parts of themselves. The intimacy of touch leads both partners to experience the healing process at a core level, leading to psychobiological, lasting change. This gradually develops into a rewiring of their love signals and pathways in creating a secure attachment bond in the relationship.

Twenty-Second Hug

All this moves the therapy toward the 20-second hug as a key way to reintegrate touch into their daily relationships. Neuroscience tells us that it takes 20 seconds of close, secure contact for oxytocin to be released by the brain and for partners to feel comforted (Grewen, Anderson, Girdler, and Light, 2003). Often, when couples have not had satisfying and intimate touch in their relationship for a long time, it is important to be directive in encouraging them to actively reintegrate touch into their relationship at home. The 20-second hug is first introduced in the couple counselling session. Couples are asked to practice within the session and to come to reintegrate close touch into their shared life. To facilitate the 20-second hug, both partners must stand up and face each other. They should be standing at a close but comfortable distance. Partners are encouraged to take a deep breath and pause for a moment, in order to be fully present with the activity. They are then invited to look into each other's eyes for a few seconds. Next, they are invited to hug each other. The therapist suggests they try and engage in the moment by relaxing into their partner's arms and staying present. The hug should be held for at least 20 seconds or more. The couple is invited to say loving words to each other.

Over time, as you repeatedly use this intervention in session, you may deepen the exercise by asking them what they notice about their body experience, their partner's body experiences, and what thoughts or emotions they are experiencing during the hug. Also, you can track therapeutic change by noticing differences in their body language as they touch. Partners often do not pay close attention to their touching behaviour, and this will be a useful activity in bringing their touching related behaviour, emotions and thoughts into the discussion.

Moving Forward

What Implications Does Touch Have for Couple Counselling?

Falling in love can last anywhere from three months to three years (Peck, 1985). Over time, however, partners' touch and talk can taper off, conflicts increase, and communication becomes more difficult. Old childhood attachment wounds, perhaps originally learned through inadequate touch, can now be brought back to life within the couple relationship. Couple counselling is about healing and repairing childhood attachment and touch wounds (Rovers, 2005). It's safe to say that most couples who come into therapy are seeking to get back to that feel-good time when they were closest and when touch was at its best. Touch techniques in couple therapy can heal childhood attachment wounds and assist couples in re-creating secure and loving attachment.

Touch in couple therapy also has the potential to expedite the process of therapy through frequent intimate engagement. Touch facilitates communication as a nonverbal form in addition to the verbal nature of therapy. Touch will increase partners' bonds and sense of connection as they get to know each other in this attuned way. Partners can also provide physical comfort for each other through the emotionally demanding process of therapy. By learning to improve their touch experiences, the couple will develop a valuable resource that can be relied on in times of distress and will serve as a protective stress reducer and healer. Touch has the potential as well to lead to a deeper spiritual connection between the couple and a more fulfilling spiritual experience with oneself. Such positive experiences with touch in the couple relationship will also be transferred to improvements in other relationships and lead to more positive beliefs about oneself, others, and the world.

Ethical Considerations

The therapist would need to be adequately trained in using touch in couple therapy. Training would involve understanding the therapeutic benefits and use of touch, knowing when it is clinically relevant to incorporate touch in the session, and knowing what constitutes appropriate and acceptable forms of touch. Above all, the use of inter-partner touch demands a secure and safe relationship between partners. When is the best time, and the best approach, to using inter-partner touch within couple counselling? Special care would need to be taken with victims of past sexual or spousal abuse, and the current sense of safety in the relationship needs to be assessed. There must also be therapeutic intent and theoretical justification for incorporating touch into a couple's treatment.

Need for Future Research

As has been previously described, there is a noteworthy lack of research specifically on inter-partner touch in couples counselling (Petrella, 2013; Machan, 2013). There is a need for research that explores the use of touch in couple counselling, the type of touch in couple counselling, the sequencing of touch as a means to repair attachment wounds, the interplay between touch and talk in couple counselling, and assessment of the different experience of touch for genders and cultures. The sequencing of touch would involve use of eye contact and healing words in the repair of wounds.

In terms of touch, we have come far from the days of Harlow's rhesus monkey experiments. We understand attachment much better; we know more about the ways of secure attachment bonds; we better understand why partners "fight for love" (Rovers, 2005)! It is time to bring the full healing power of touch to bear for partners within couple counselling sessions.

Note

1. "Inter-Partner Touch in Couple Counselling: Theory and Emerging Practice," by Cassandra Petrella and Martin Rovers, was previously published in the *Journal of Counselling and Spirituality, 33*(1) 2014, pp. 93–109. Permission to reprint granted.

References

Ainsworth, M. D. (1989). *Patterns of attachment: A psychological study of the strange situation*. New York: Eribaum.

Bowlby, J. (1969). *Attachment and loss: Volume 1. Attachment*. London: The Hogarth Press and the Institute of Psycho-Analysis.

Berg-Cross, L. (2001). *Couples therapy*. New York: The Haworth Press.

Cholewiak, R. W., and Collins, A. A. (1991). Sensory and physiological bases of touch. In M.A.Heller & W. Schiff, *The Psychology of Touch*. New Jersey: Lawrence Erlbaum.

Coan, J. A., Schaefer, H. S., and Davidson, R. (2006). Social regulation of the neural response to threat. *Psychological Science, 17*(12), 1032–1039.

Dadds, M. R., Allen, J. L., Oliver, B. R., Faulkner, N., Legge, K., Moul, C., and Scott, S. (2012). Love, eye contact and the developmental origins of empathy v. psychopathy. *British Journal of Psychiatry, 200*, 191–196.

Engebretson, J., and Wardell, D. W. (2012). Energy therapies: Focus on spirituality. *Explore: The Journal of Science and Healing, 8*(6), 353–359.

Feeney, J. A. (1998). Adult attachment and relationship-centered anxiety: Responses to physical and emotional distancing. In J. A. Simpson and W. S. Rholes (Eds.), *Attachment theory and close relationships*. New York: Guildford Press.

Feeney, J. A., and Noller, P. (1991). Attachment style and verbal descriptions of romantic partners. *Journal of Social and Personal Relationships, 8*, 187–215.

Field, T. (2002). Infants' need for touch. *Human Development, 157*, 1–4.

Field, T., Scafidi, F., and Schanberg, S. (1987). Massage of preterm newborns to improve growth and development. *Pediatric Nursing, 13,* 385–387.

Gerber, R. (1996). *Vibrational medicine: New choices for healing ourselves.* Santa Fe, NM: Bear and Company.

Grewen, K. M., Anderson, B. J., Girdler, S. S., and Light, K.C. (2003). Warm partner contact is related to lower cardiovascular reactivity. *Behavioural Medicine, 29,* 123–130

Grewen, K. M., Girdler, S. S., Amico, J., and Light, K.C. (2005). Effects of partner support on resting oxytocin, cortisol, norepinephrine, and blood pressure before and after warm partner contact. *Psychosomatic Medicine, 67*(4), 531–538

Harlow, H. (1958). The nature of love. *American Psychologist, 13,* 673–685.

Johnson, S.M. (2008). *Hold me tight: Seven conversations for a lifetime of love.* New York: Little, Brown and Company.

Levitan, A., and Johnson,J. (1986). The role of touch in healing and hypnotherapy. *American Journal of Clinical Hypnosis, 28*(4), 218–223.

Machan, M. (2013). *Touch: A Literature Review of the Biology and the Philosophy of Touch.* Unpublished master's research paper, Saint Paul University, Ottawa, Ontario, Canada.

Maltz,W. (Producer). (1995). Relearning touch: Healing techniques for couples [DVD].

Masters, W., and Johnson, V. (1970). *Human Sexual Inadequacy.* Boston: Little Brown.

Mikulincer, M., and Shaver, P. R. (2010). Attachment in adulthood: Structure, dynamics and change. New York: The Guilford Press.

Montagu, A. (1971). *Touching: The human significance of the skin.* New York: Columbia University Press.

Montagu, A. (1978). *Touching.* San Francisco: Harper and Row.

OECD. (2007). *Understanding the Brain: The Birth of a Learning Science.* OECD

Peck, M. S. (1985). *The Road Less Travelled.* New York, NY: Simon and Schuster.

Petrella, C. (2013). *Inter-Partner Touch in Couple Counselling: A Review of the Literature and Future Implications.* Unpublished master's research paper, Saint Paul University, Ottawa, Ontario, Canada.

Rovers, M. W. (2005). *Healing the Wounds in Couple Relationships.* Ottawa: Novalis

Rovers, M. W. and Kocum, L. (2010) Development of the Holistic Model of Spirituality. *Journal of Spirituality in Mental Health,* 12: 1, 2–24.

Rovers, M.W. and Petrella, C. (2013). *Inter-Partner Touch Scale.* Ottawa.

Shapiro, D. L., and Levendosky, A. A. (1999). Adolescent survivors of childhood sexual abuse: The mediating role of attachment style and coping in psychological and interpersonal functioning. *Child Abuse and Neglect, 23,* 1175–1191.

Solomon, M., and Tatkin, S. (2011). *Love and war in intimate relationships: Connection, disconnection and mutual regulations in couple therapy.* New York: W. W. Norton and Company.

Spitz, R. A. (1945). An inquiry into the genesis of psychiatric conditions in early childhood. *Psychoanalytic Study of the Child, 1*, 53–74.

Tronick (2003). *Still Face Experiment*. Retrieved from www.youtube.com/watch?v=apzXGEbZht0

Weber, R. (1990). A philosophical perspective on touch. In K. E. Barnard and T. Berry Brazelton (Eds.), *Touch: The Foundation of Experience*. Madison, Connecticut: International Universities Press.

PART III

THE PRACTICE OF TOUCH IN A SPECTRUM OF PRACTICE

Chapter VIII

The Practice of the Laying on of Hands in Islamic Spiritual Care

Nazila Isgandarova

The practice of the laying on of hands represents more traditional beliefs and practices in Islamic spiritual care, which usually provides information, advice, encouragement, and instruction to the client by using various religious and/or spirituality-based texts. Nevertheless, despite the fact that recently Muslim spiritual caregivers actively have begun to use the knowledge and skills of psychotherapy, such as therapeutic listening, reflection, and decision-making based on the social sciences (i.e., psychology, psychotherapy, ethics, family studies, etc.), some of them also rely on a unique and controversial aspect of traditional healing practices, including the practice of the laying on of hands in Islamic spiritual care. This may be explained with the fact that since the time of the Prophet, the practice of the laying on of hands performed a particular function for the person on whom it was performed. Early Muslims strongly believed that not only the words but also the ritual of the laying on of hands carried the potential for healing and deliverance. This belief is still in the hearts of many Muslims who prefer traditional healing practices. If we take into consideration that Islamic spiritual care reflects various aspects of existing beliefs and practices in society, in Muslim societies and communities where these practices are allowed, Muslim spiritual caregivers are expected to include them in their caregiving.

In this chapter, I give a description of the practice of laying on of hands in the context of Islamic spiritual care. In the section on the literature review, I discuss the practice of the laying on of hands and its origins. I also briefly explore the similarities and differences of Islamic traditional healing practices and describe how the differences among Muslims affect their understanding of healing practices. The second part of this chapter describes the practice of the laying on of hands and represents it as a traditional healing practice in Islamic spiritual care. I discuss traditional healing as an alternative treatment to conventional medicine and argue that more research is needed on the referral process in traditional healing practices, the nature of the problems referred, the effectiveness of the referral process, the relationship between specific healing traditions in Islam and other forms of alternative care. The third part of this chapter explores the ethics of the laying on of hands in order to minimize the risks associated with it. The main questions that guide this paper are "How is the laying on of hands used in Islamic spiritual care in order to contribute to the physical, spiritual, emotional, and mental health of Muslims?" and "How does Islamic spiritual care respond to the risks associated with the laying on of hands as a traditional practice?"

Literature Review

The practice of the laying on of hands in Islamic traditional healing practices is not to be assumed as specifically Islamic because, like other aspects of traditional healing, the laying on of hands existed before Islam. For example, in ancient Israel, many prophets such as Jacob, Aaron, and others, blessed their people with their hands crossed on their heads (Genesis, 48: 14–16; Lev. 9:22, Psalms, 139:4–6), or by doing so, they indicated that the transfer of authority from one person to another had been completed (Deut. 34:9). The Gospel also introduces many narratives where Jesus blesses his disciples or children either by putting his hands on their heads or lifting up his hands for benediction (Mark, 10:16; Luke, 24:50). Despite the fact that the practice of laying on of hands existed before Islam, this practice is mainly accepted as a prophetic tradition in Islam. Therefore, this practice represents the ceremonial aspect and application of Islamic healing traditions in Islamic spiritual care. Overall, the practice of the laying on of hands is performed together with other healing rituals and ceremonies that have been recorded or endorsed in the key written and oral texts of Muslims, especially in *hadith* literature (a collection of narratives about the practices of the Prophet Muhammad).

The earliest records of the laying on of hands as a sacramental and healing practice were written by Anas ibn Malik in the eighth century. *The Prophetic Medicine* by Ibn Qayyim al-Jauziyya (d. 1350) became more popular among Muslims with regard to healing through application of prophetic tradition. These books[1] extensively report miracles performed by the Prophet Muhammad through the laying on of hands, and can give the impression of unbelievable power possessed by the Prophet Muhammad. Some classical Muslim scholars also believed that the practice of laying one or both hands on the afflicted person brings countless benefits. In the Islamic tradition, healing particularly refers to spiritual healing or *shifa*, and does not mean a medical cure or *mualija* in Arabic. The Muslim tradition of spiritual healing is based on the recognition of the effect of spiritual health on the physical body, which is seen as a mere receptacle for the spirit. Classical Qur'anic scholars such as Abu Ja'far Muhammad ibn Jarir at-Tabari (839–923) and Ismail ibn Kathir (1301–1373) interpreted *shifa* as a remedy for spiritual diseases of the heart. These diseases are defined as greed, haughtiness, and selfishness, and healing them "through supplications, prayers, and fasting is a well-established tradition among Muslims" (Iqbal, 1998, p. 3). Contemporary literature on spiritual care also differentiates between spiritual healing and a medical cure. For example, Michael Lerner (1996), an author of complementary therapy programs and spiritual and psychological approaches, points out that if the goal of medicine is "curing," then the goal of mind-body effort should be "healing," which comes from the "inner resources" of the person.

Some works by other classical Muslim scholars questioned healing through sacramental acts and tried to bring to the attention healing through physical order.[2] More contemporary Muslim scholarship reports a strong belief among Muslims in healing through sacramental acts (Rahman, 1987; Ghali, 2014). Both classical and contemporary Muslim scholarship extensively relies on the *hadith* literature when they describe the belief in sacramental acts such as the laying on of hands. This is because, with regard to the laying on of hands, there is no specific reference to it in the Qur'an. I agree with the idea that the *hadith* literature developed a rich tradition of seeking cure for physical and mental illnesses, and it describes how the Prophet and his companions practiced reading prayers, laying on of hands, asking for a cure from God for an afflicted person. The *hadith* literature also narrates how the Prophet's companions witnessed him practicing the laying on of his hands on the chest or forehead of a sick person and reciting Qur'anic verses or invocations for healing. For the companions, this act of the Prophet indicated the act of wishing the sick person to heal faster or carried a message of a personal blessing. For example, Sahl ibn Sa'eed, one of the Prophet's companions, narrated that during the battle of Khaybar, the Prophet (S) spat in Ali's sore eyes and he recovered as if he had never suffered any pain (Sahih al-Bukhari, 1984, p. 20). On another occasion, the aunt of as-Sa'ib ibn Yazid took him to the Prophet and asked him to cure him. The Prophet rubbed the child's head and prayed God to bless him. He also performed wudhu' (ablution) and asked the child to drink from the water of his wudhu' (Sahih al-Bukhari, p. 20). In another incident, Abu Abdurrahman al-Yamani, who was *tabiun* (one of the greatest scholars of the generation that followed the age of the Companions), reported that "Whenever an insane person came to God's Prophet, he was cured as soon as God's Prophet put his hand on the chest of the ill person..." (Sahih al-Bukhari, 1984, p. 20).

The literature on the traditional spiritual care and medicine (Crapanzano, 1973; Erez, 1981; Al-Krenawi and Graham, 1997; Al-Issa, 2000; Adib, 2004; Dole, 2004; Abdulla, 2007; Yucel, 2008; El-Islam, 2008; Isgandarova, 2011) does not specifically indicate plainly whether the practice of the laying on of hands was (and is) a sacramental act. Although some authors (Isgandarova, 2011) do not emphasize employing this practice extensively in Islamic spritual care, they argue that the caregivers should not ignore the belief of clients who believe in the possibility of healing through rituals, including the practice of the laying on of hands.

The literature also sheds light on various practices of the laying on of hands, including prayer associated with it and its benefits. According to Isgandarova (2011), denominational variance affects the practice of Islamic spiritual care because the Shi'a and Sunni branches of Islam have some different views on healing practices among Muslims. For example, historical and contemporary practices of traditional healing indicate that Shi'a and Sunni branches of Islam may have extremely different views on traditional healing practices; if the practices of the companions of the Prophet Muhammad play an important role in Sunni traditional healing, including the laying on of hands, then the Twelve Imams for Twelve Shiites of Shi'as and the Agha Khan in the Ismaili tradition in Shi'a healing practices also hold special meaning. In the Twelve Shiites tradition, the medicine of the imams (Tibb *al-A'imma*), dating back to the eighth Imam Ali al-Rida (d. 818), is also very famous (Ghali, 2014). I explain the heterogenous practices with an open and flexible interpretative system of Islam that allows local and theological contexts to coexist within the Islamic tradition. Furthermore, socio-historical predicaments and local customs and practices of a given society (*adat* or *'urf*) also add more variations to the practice of the laying on of hands in Islamic spiritual care.

In sum, the literature on the practice of the laying on of hands does not specifically explore whether this practice has a place for healing in Islamic spiritual care. The lack of clarity around it prevents us from making a definitive conclusion. Nevertheless, the review of traditional sources and more recent literature on this subject suggests that the practice of laying on of hands cannot simply be overlooked because many Muslims do not take the stories about the healing through the laying on of hands as mythology. Although this practice integrates some ancient practices of healing (or folklore) before Islam, it became Islamic or accepted as compatible with the Islamic tradition through narratives that attest to the healing power of the laying on of hands by the Prophet Muhammad and other Muslim saints.

A Place for the Laying on of Hands in Islamic Spiritual Care

The practice of the laying on of hands is not widely practiced in Islamic spiritual care in North America. Nevertheless, as a sacramental act of healing, the practice of the laying on of hands in Islamic spiritual care represents a traditional healing practice within the history of Islam. This practice is especially widely employed during the application of herbs and medications known as *unani*, use of exorcism or *ruqya* (in Arabic this means a charm, spell, or incantation), pilgrimages to holy sites and shrines, and so on.

The practice of *ruqya* is specifically important to mention here because of its reliance on the practice of the laying on of hands while reciting the Qur'an or making *du`a* (supplication) to God for the patient. All these acts during the process of *ruqya*, particularly the laying on of hands, perform a function of a formal declaration of transmitting healing to the afflicted person, or at least removing the physical, emotional, and spiritual disease from him/her. Some Muslims even consider that the laying on of hands make the Qur'anic supplications more effective. Although I have never witnessed or heard any Muslim spiritual caregiver in North America perform this practice in a health care setting, I have heard that Muslim spiritual caregivers in traditional settings still widely perform this ritual or have been asked to employ it. The controversial aspect of this practice in Islamic spiritual care can be explained by its application in the practice of exorcism, or freeing the person who is assumed to be possessed by evil spirits—unseen, gendered creatures that are made of fire. According to folk belief, bad spirits inflict people with maladies and cause harm to their well-being. The afflicted person may be possessed either through sorcery or from the spirit's desire for another person (Abdulla, 2007). For a long period of time, the afflicted person may not know that he/she has been possessed (Crapanzano, 1973).

In my experience, I have seen many Muslim spiritual caregivers in traditional settings, who are well trained in *ruqya*, perform an exorcism to free the afflicted person. I also found the process of *ruqya* to be a lengthy process that requires a good memory of the Qur'anic verses. Nevertheless, during my interviews with some imams and Muslim spiritual caregivers in 2011, I witnessed that a simple process is also prescribed, which includes reciting the opening chapter (*surah al-Fatihah*) seven times, because Muslims in general view the Qur'an as a "book of healing" (Isgandarova, 2011). The recitation of the first chapter of the Qur'an is usually followed with the recitation of *ayat al-Kursi* from the second chapter, which includes the most powerful verses in the Qur'an, and *surah al-Falaq* (the Chapter of Dawn or Daybreak) and *surah al-Naas* (the Chapter of Humankind). During the recitation process, the caregiver may lay the hand on the person who requests this ritual. If the afflicted person recites these prayers himself/herself, after the recitation has been completed, he/she blows on their hands and rubs them over the body.

In contemporary Islamic spiritual care in a health care setting, the practice of the laying on of hands is usually viewed as something suspicious and harmful. Many Muslim spiritual caregivers usually do not accept the laying on of hands as a mainstream healing technique. However, they also do not reject or undermine the stories of healing through it. This could be explained by the fact that the theological foundation of the sacramental acts, such as laying on of hands, reading the sacred verses over water and drinking from it, comes from the prophetic tradition. For example, the *hadith* literature points out that the Prophet Muhammad himself used to read the prescribed prayers from the Qur'an, blew his breath into his hands, and wiped his hands over his face, or put his hands on the afflicted person. It is generally accepted that every Muslim can practice on himself/herself or those who ask them to do so, if they feel comfortable. Therefore, it is a common belief among many Muslims that the Prophet Muhammad's means of healing were not only through words but also by touching the sick person with his hand.

Nevertheless, the Qur'an does not report only stories of healing that involved the practice of the laying on of hands. In the chapter al-Maida (Q. 5: 6), rubbing of the head, face and hands is mentioned in the context of *wu'du* (ablution with water) or *tayammum* (ablution using sand or dust). Although there are some narratives that describe the practice of the laying on of hands by the Prophet Muhammad, a majority of the *hadith* include narratives on how the Prophet Muhammad advised his companions to take preventative measures against illnesses, such as leprosy and epidemics, by observing proper hygiene and forbidding alcohol and suicide. Therefore, a more general approach in Islamic spiritual care today emphasizes proper medical treatment and professional counselling for mental and physical health problems (Isgandarova, 2011). Muslim spiritual caregivers who support this approach believe that the prophetic tradition encourages Muslims to seek various ways for the healing and treatment of diseases. For example, Prophet Muhammad said, "There is no disease that God has sent down, except that He has also sent down its treatment" (Sahih al-Bukhari, p. 1232). However, why do Muslims sometimes prefer to use Islamic traditional healing, such as the practice of the laying on of hands, not only for treating neurotic symptoms and minor ailments but also for severe mental or physical illness? Why have Muslim spiritual caregivers who perform the practice of the laying on of hands become attractive potential sources of assistance with mental health problems?

When I pose these questions, I do not imply that Islamic spiritual care should ignore various ways of healing or should reject the idea of the practice of the laying on of hands since it does not fit with contemporary medicine. I am aware that some studies also suggest that mainstream mental health services downplay the idea that traditional healing practices may constitute a significant part of the healing process for Muslims. This is because they see traditional healing practices as contradictory to current mental health theories and practices. In the context of Aboriginal healing practices, B. Hames Waldrum (1997) criticizes a general approach of conventional medicine to traditional healing by asking:

> Is this because the healing itself is ineffective? Is it possible that science currently lacks the tools (and the inclination) to see "traditional healing"? How do healers view the issue of efficacy? What are their goals when they undertake healing and what measures (if any) do they employ to determine success? Is efficacy, as science understands it even an issue for traditional healers? Is the whole idea of questioning efficacy and developing evaluation programs even necessary? Is it possible that to undertake these we are violating the basic principles of the healing itself? Can traditional healing ever be understood by the dominant biomedical system? (pp. 26–28)

In the Islamic context, it is true that Muslims encounter, engage, and exchange ideas and best practices in the health care system; some of them may find the practitioners of conventional medicine are ignorant of their beliefs and practices, such as the laying on of hands in traditional healing. They may prefer to see more traditional aspects of healing, such as the practice of the laying on of hands and reading the Qur'anic verses, integrated in Islamic spiritual care together with counselling, information giving, and the like. In situations where caregivers ignore or undermine a traditional perspective of healing, some Muslim patients may feel that their needs are not met. In order to avoid this tension, it is important to remember that the primary aim of Islamic spiritual care is sustaining and supporting the sick person through reconciliation, and offering necessary rituals and traditions of healing and forgiveness.

Ethics of the Laying on of Hands in Islamic Spiritual Care

The practice of the laying on of hands in Islamic spiritual care is not a simple matter. It involves a whole range of ethical issues. Muslim spiritual caregivers are accepted as mental health professionals or as leaders and role models of their communities, and are perceived to be significantly more developed in interpersonal skills, including warmth, caring, and stability. As a general requirement, those who claim to use this title are required to live according to the prophetic tradition in order to be effective in their spiritual care to others. This expectation is further enhanced with the works[3] of Muslim scholars, who laid the foundation of ethics in Islamic spiritual care and put an emphasis on practicing sacramental acts such as the laying on of hands properly and ethically.

The following subsections highlight the ethics of the laying on of hands in Islamic spiritual care and remind that the functions of this practice are mainly healing, sustaining, guiding, and reconciling.

Ethics of the *Tawhid* Principle

The primary ethical framework of Islamic spiritual care has been developed around the *tawhid* (oneness and unity of God) principle. The Qur'an specifically emphasizes the belief that healing comes only from God. However, the *hadith* literature is not explicitly clear whether the Prophet's companions believed that the Prophet had a special healing power, or if he was acting as a medium in transferring divine healing to the sick person when he used the practice of the laying on of hands. I assume that a majority of the companions did not ascribe such a healing power to the Prophet himself because attributing to the Prophet Muhammad or the diseased saints the power of divine healing is generally viewed as *bid'at* (innovation) or *shirk* (ascribing something to God). Later on, some generations of Muslims viewed the prophetic practice of the laying on of hands as a symbolic and formal way of invoking blessing from God. For them, the Prophet Muhammad was an intercessor between God and humans; some of them even ascribed a divine power to the Prophet, and his practice of the laying on of hands signified his divine authority. I have also seen many Muslims who continue to believe that the Prophet Muhammad transmitted the sacred power or healing to the person when he put his hands on them. They also believe that some saints possessed this divine power in their hands and could heal people through the practice of the laying on of hands. Nevertheless, it is also generally accepted that no one can represent God's healing power. As al-Issa (2000) points out, "both exorcism and magical practices seem to have been allowed in Islamic societies, as long as they are practiced in the name of God to help people . . . rather than inflict harm on them in collaboration with the devil" (p. 64).

Therefore, the healing process through the laying on of hands may be viewed as something unknown or mysterious. However, in our contemporary practice of Islamic spiritual care, there is a tendency to explain the healing process by evidence-based knowledge and not leave it to human fantasy and imagination. Our experience of divine intervention or healing is personal because our relationship with God is not hierarchical; instead it is at the centre of our lives, and we feel the effects of this relationship on a personal level. Healing happens within the person, but the origins of healing stem from God, and it is a process that happens beyond the human intervention.

Ethics of Physical and Spiritual Purification

Another ethical challenge of practicing the laying on of hands in Islamic spiritual care is with the difficulty of applying all the strict guidelines of traditional healing practices with respect to the laying on of hands. For example, most Muslim scholars generally state that the practice of the laying on of hands and also saying supplications (*du 'a*) must be preceded with a proper physical and spiritual preparation. The practical guideline for this kind of preparation is summarized in the following:

1. Before the healing practice with the client, it is necessary to take ablution (*wudhu'*) and ensure that the area and clothes are clean;
2. There should be a strong faith that the practice will be responded to, either in this world or in another form in the Hereafter;
3. A level of optimism of the client that the healing practice will be accepted must be achieved;
4. Performing healing practices may be more beneficial during auspicious times of the day, week, month, or year, such as during times of distress, when travelling, on Fridays, during Ramadan, on holy nights, i.e., *Laylat al-Qadr* (the Night of Power, which is believed to be the 27th day of Ramadan), and after obligatory prayers or in certain places such as in the Ka'bah in Makkah, the Prophet's Mosque in Madinah, or Masjid al-Aqsa in Jerusalem;
5. Healing practice must take place in a quiet place where no one is distracted;
6. Finally, acquiring a *halal* (lawful) livelihood and consuming *halal* food is key to getting a positive result.

Ethics of Providing Proper Care

Muslim spiritual caregivers appear to be an important source of mental health assistance. However, some Muslim spiritual caregivers, especially in traditional settings, have significant limitations in identifying emotional distress or suicide risk. If the person poses a danger to others, they need to use the evaluation of diseases and referral skills as additional sources of assistance for certain types of mental health problems. They may help the client/patient to decrease his/her symptoms using traditional healing techniques, including the practice of the laying on of hands. However, persons who are psychotic, suicidal, have addiction or substance abuse problems, and those who are severely depressed and confused should be referred to other professionals, such as counsellors, social workers, psychiatrists, or psychologists, for additional help. Ignorance of ethics of proper traditional healing practices in Islamic spiritual care can lead to negative consequences. The practice of the laying on of hands as traditional healing may only seem effective in addressing particular needs of the afflicted person for a short period of time.

One of the ethical challenges with regard to the practice of the laying on of hands is the attempt to diagnose mental and emotional problems or prescribe certain things during the process of the laying on of hands. I have seen some Muslim spiritual caregivers who recommend reciting verses from the Qur'an and asking their clients to blow their breath to the affected area. Some of them may even recommend carrying amulets or *muska* in order to keep evil spirits away from the person. However, my suggestion to these Muslim spiritual caregivers who use the traditional healing practices, including the practice of the laying on of hands, is to pay attention to whether their approach leads to adverse psychological and physical consequences. They should remember that the Prophet emphasized correct methods of healing and curing for serious problems and encouraged his followers to seek professional help.

Conclusion

The practice of the laying on of hands is a sacramental traditional healing practice in Islamic spiritual care. However, it does not replace mainstream healing methods, such as counselling and advice-giving. Nevertheless, in traditional settings, where access to a health care system is not easy and fast, the laying on of hands is accepted as an important healing technique. This practice carries several potential meanings. First, applying traditional methods of healing such as the laying on of hands refers to religious and spiritual activities that are believed to hasten recovery. Second, it refers to the practice of Muslims who prefer to seek care through both Islamic traditional healing and conventional medicine in the hopes that they may benefit from both. Third, the laying on of hands as a traditional healing practice is a socially situated activity, which implies that the practice extends beyond the person and reaches the family, community, and society. Fourth, in some situations, traditional healing practices replace mainstream treatment and may even be more effective in addressing particular needs in a short period of time. This is an especially widespread phenomenon in developing countries, where medical professionals may fail to treat mental disorders.

The practice of the laying on of hands in Islamic spiritual care cannot be labeled as against Islamic belief because the *hadith* literature also indicates that the Prophet used to practice it. However, according to the Prophetic tradition, healing through the practice of the laying on of hands does not refer to any process or activity that has a particular beginning or end; instead, healing is a lifelong process or journey. Moreover, it encompasses all four elements of a whole person: spirit, emotion, mind, and body. In Islamic spiritual care, an absence of healing is assumed to be caused by a fragmented sense of self. Therefore, many Muslims do not distinguish between one's mind, emotion, spirit, and body. True healing happens when there is a unification of these four important aspects of a human being. The primary responsibility of a Muslim spiritual caregiver is to help the person to achieve this balance between mind, emotion, spirit, and body.

Nevertheless, Islamic spiritual care needs more research to explore answers to questions such as: Who refers patients for this care, what kind of problems are referred, and how well does that referral process work? Is the laying on of hands as a sacramental and traditional healing practice efficacious? In what ways and under what circumstances is the laying on of hands effective? How do health care systems justify and support the provision of traditional healing to Muslims? What parallels are evident between specific healing traditions in Islam and other forms of alternative care? I strongly believe that any research in this field is important because of the significance given to the laying on of hands and other forms of traditional and sacramental healing among some Muslims.

Notes

1. Ibn al-Sunni al-Dinawari (d. 974), Abu Nu'aym al-Isfahani (d. 1038), Abu al-'Abbas al-Nasafi (d. 1041), Ibn al-Jawzi (d. 1200), Diya' al-Din al-Maqdisi (d. 1245), Shams al-Din al-Ba'li (d. 1309), Shams al-Din al-Dhahabi (d. 1348), Ibn al-Qayyim (d. 1350), Ibn Muflih (d. 1362), Jalal al-Din al-Surramari (d. 1374), Jalal al-Din al-Suyuti (d. 1505), Ibn Tulun al-Dimashqi (d. 1546), and Shihab al-Din al-Qalyubi (d. 1658). Nevertheless, among these books only *Beneficial Selections from Medicine* by Abu'l-Faraj ibn al-Jawzi (d. 1200), a Hanbali jurist, *Prophetic Medicine* by al-Dhahabi (d. 1348), *Al-Tibb al-Nabawiyy* or *Kitab al-Mansuri, Kitab al-Hawi fi al-tibb* (*The Comprehensive Book on Medicine*), and *Kitāb al-Tibb al-Ruhani (The Book of Spiritual Physick)* Abu Bakr Muhammad ibn Zakariya al-Razi (or Rhazes) (d. 925 or 932), Abu 'Ali al-Husayn ibn 'Abd Allah ibn Al-Hasan ibn Ali ibn Sina (980–1037), *The Book of Healing* and *Al-Qanun fi al-Tibb* (The Canon of Medicine) by Abu Ali Ibn Sina (d. 1037), etc., convey more rational approaches to healing.
2. Nevertheless, some, i.e., Ibn Sina, also did not ignore healing through traditional healing practices such as prayer. According to Ghaly (2014), Muslim physicians such as Abd al-Latif al-Baghdadi (d. 1231), Ibn Tarkhan (d. 1320), Mahdi al-Sanawbari (d. 1412), and Ibrahim al-Azraq (d. 1485) were also interested in healing and treatment though prophetic tradition.
3. Examples of these works are Imam al-Ghazali's (1058–1111) *Ihya Ulum ad-Din* (The Revival of Religious Science), Imam Nawawi's (1234–1278) *Riyadh as-Saliheen* (Gardens of the Righteous), Said Nursi's (1877–1960) *The Letters* and *the Flashes,* and Wahbi Zuhayli's *The Encyclopedia of Islamic Jurisprudence.*

References

Abdulla, S. (2007). Islam and counseling: models of practice in Muslim communal life. *Journal of Pastoral Counseling, 42*: 42–55.

Adib, S. M. (2004). From the biomedical model to the Islamic alternative: A brief overview of medical practices in the contemporary Arab world. *Social Science Medicine, 58:* 697–702.

Al-Issa, I. (2000). Mental illness in medieval Islamic society. In I. Al-Issa Ed.), *Al-Junun, Mental Illness in the Islamic World* (p. 64). Madison: International Universities Press.

Al-Jawziyyah, Ibn Al-Qayyim. (1999). *Healing with the Medicine of the Prophet.* Trans. Jalal Abu Al-Rab. (Ed. Abdul R. Abdullah). Riyadh: Darussalam Publications.

Al-Krenawi, A. and Graham, J. (1997). Spirit possession and exorcism in the treatment of a Bedouin psychiatric patient. *Clinical Social Work Journal, 25:* 211–222.

Crapanzano, V. (1973). *The hamadsha. A study in Moroccan ethnopsychiatry.* Berkley: University of California Press.

Dole, C. (2004). In the shadow of medicine and modernity: medical integration and secular histories of religious healing in Turkey. *Culture, Medicine, and Psychiatry, 28*: 255–280.

El-Islam, M. F. (2009). Islamic religion and traditional healers' contribution to mental health and well-being. In M. Incayawar, R. Wintrob, L. Bouchard and G. Bartocci (Eds.), *Psychiatrists and traditional healers: Unwitting partners in global mental health* (pp. 197–205). Oxford: Willey-Blackwell.

Erez, S. (1981). Ottoman medicine viewed through Western eyes. *Hexagon Roche, 9*(3): 1–4.

Ghaly, M. (2014). Prophetic medicine. In C. Fitzpatrick, and A. Walker (Eds.), *Muhammad in history, thought, and culture: An encyclopedia of the Prophet of God* (2 vols.) (Vol. II, pp. 502–506). Santa Barbara: ABC-CLIO.

Isgandarova, N. (2011). *Effective Islamic spiritual care: Foundations and practices of imams and other Muslim spiritual caregivers.* Doctor of Ministry thesis, Waterloo: Wilfrid Laurier University.

Iqbal, M. (1998). Islamic medicine: The tradition of spiritual healing. *Science & Spirit, 9*(4).

Lerner, M. (1996). *Choices in healing: Integrating the best of conventional and complementary approaches*. Cambridge, MA: MIT Press.

Rahman, F. (1987). *Health and medicine in the Islamic tradition: Change and identity*. New York: Crossroad.

Sahih al-Bukhari. (1984). Translated by M. Muhsin Khan, New Delhi, India: Kitab Bhaban.

Smith, M. (1994). *Rabi'a: The life and works of Rabi'a and other women mystics in Islam*. Oxford: Oneworld Publications.

Waldrum, B. H. (1997). But does it work? Traditional healing and issues of efficacy and evaluation." In Widening the circle: Collaborative research in mental health promotion in Native communities. Proceedings of the Conference, September 26–28, 1997. Montreal.

Yesilada, Y. (2011). Contribution of traditional medicine in the healthcare system of the Middle East. *Chinese Journal of Integrative Medicine, 17*(2): 95–98.

Yucel, S. (2008). *The effects of prayer on Muslim patients' well-being*. D.Min. thesis. Boston University School of Theology.

Chapter IX

An Angel in My Pocket: Touch, Sacred Objects, and Spiritual Coping

Linda Mayorga Miller

Touch in Hospital Chaplaincy

There are few places where a person is so entirely at the mercy of others as a health care setting. Patients are not allowed to wear their own clothing. Meal times, bedtimes, medication, treatment, tests: everything is organized for the convenience of the system. Hospital patients do not even have control over their own bodies. They may have the legal right to refuse medical procedures, but a desire to be healthy essentially nullifies that choice. A patient once told me that a diagnosis of cancer was like getting on the wrong train in a foreign country where she did not know the language. Although it is not logical, in these moments it is easy for us to believe that no one has ever experienced such a frustrating lack of control over his or her own life. Loss of control often engenders feelings of anger, fear, and helplessness, and is a key factor contributing to the experience of stress (Hood, Spilka, Hunsberger and Gorsuch, 1996; Pargament, Koenig and Perez, 2000).

In addition, serious illness is an isolating experience. It physically separates us from our normal circles of support, because we are bed-bound or in hospital. It psychologically and emotionally isolates us because, as sympathetic as friends and family may be, they are not walking in our shoes (Janet Hatanaka, personal conversation about her PhD in nursing research findings, February 7, 2006).

Touch offers a physical connection with another human being, a very basic, solid confirmation that we are not alone in our suffering. A positive moment of heartfelt physical connection in a time of distress can have a profound and long-lasting beneficial effect. It is a physical connection with something solid. It can be a point of grounding, not just because of the physical connection with another person, although that is significant; but touch may also serve as the modality for connection with a greater reality. The right touch in a crucial moment of spiritual darkness can reconnect us to humanity or help us feel linked to a universal source of love. It can even be perceived as a manifestation of God's loving presence.

Touch works as a therapeutic intervention for the spiritual care provider because touch works as a *spiritual coping strategy* for the client. As is the case with all therapeutic interventions, touch should not be used without a spiritual assessment to determine if it will be helpful and appropriate (Hodge, 2003). This is vitally important because touch, even something as simple as a hand on a shoulder, may destroy the therapeutic relationship or even cause harm to the client if, for example, it is unwanted and/or stirs up memories of injury or abuse.

This chapter stems from my research on the use and touching of sacred objects within spiritual care practice (Miller, 2008). The insights from literature and my research are combined to review the impact of touch and touching sacred objects within spiritual care (Miller, 2008: Miller, Gall, and Corbeil, 2011) and ways that spiritual care practitioners and chaplains might understand touch in their helping relationship.

Spiritual Coping

There are many reasons people engage in spiritual practices. They differ depending on the situation, context, need, and time of life. Specifically, stress, suffering, worry, and fear tend to open the door to spiritual coping strategies. Kenneth Pargament (1997) has identified two common and significant reasons people turn to spiritual coping, particularly in times of stress.

Relatively Available. When religion and/or spirituality are a way of life, rather than a part of life, they hold a central place, influencing many aspects of living. In times of stress the very religious or spiritual would easily turn to pre-existing theological beliefs and spiritual practices to cope (Samson and Zerter, 2004). For others, religion and spirituality are compartmentalized. They play a small part or no part at all in global meaning, so they would play an equally small part in coping.

Relatively Compelling. Serious situations which push an individual to the limits of her/his resources and abilities bring human limitations to the forefront. In such situations coping methods which access a greater reality are more compelling than secular strategies (Pargament, 1997).

In the end, the choice to use a religious or spiritual coping strategy is made on a cognitive as well as an emotional level. It is a compelling choice when, because of the individual's preexisting spiritual framework, it makes sense, and because of dire circumstances, it feels right.

Locus of Control

Locus of control refers to the individual's perception of whether events in her life are due primarily to chance, fate or God, versus being the result of her own choices and actions. Locus of control has been identified as a significant factor in spiritual coping. However, research in this area reveals that it is a surprisingly complex factor (Levin, 2001). Jackson and Coursey (1988) found that both perceived high God control and high personal control are related to various dimensions of positive coping. They present the argument that through prayer an individual may feel that she influences God to intervene on her behalf and in this way gains a sense of control over events which are otherwise beyond her influence. This contradicts the previously long-held belief that spiritual coping techniques are antithetical to personal efficacy because they place control externally with God. The concept that spirituality is more than a passive or defensive coping strategy has been supported elsewhere (Pargament and Park, 1995). Ellison and Taylor (1996) hypothesized that people often turn to spiritual practices specifically to gain control over the negative emotions, which arise in reaction to events that lie outside the realm of personal problem solving capabilities, such as serious illness, disabilities, chronic pain or bereavement. While Ellison and Taylor recognized that people were trying to gain control over negative emotions, they didn't explore the effectiveness of spiritual coping. Others found that spiritual coping has shown itself to be a complex construct that "adds a unique component to the prediction of adjustment to stressful life events that cannot be accounted for by other established predictors" (Tix and Frazier,

1998, p. 420). Pargament and his colleagues posit that this unique component is the sacred (Pargament, Magyar-Russell and Murray-Swank, 2005).

The Spiritual Dimension of Touch

The spiritual dimension of touch receives its authority from a transcendent reality which is perceived and experienced by the client. Theologians have long agreed that transcendent reality is wholly outside the mundane world, yet underlies all life experience (Dupré, 2000). According to Dupré, believers consider the transcendent to be an absolute eternal reality, while the physical world is transient and mutable. Psychologist Kenneth Pargament (2007) agrees, arguing that concepts of transcendence (e.g., Heaven, love, Creator, spirits of deceased loved ones) reside at the core of sacredness. Pargament goes on to explain that any object, person, place, time, space, attribute, or activity can be sacralized through its association with this sacred core. These things become invested with divine character or qualities (as opposed to being perceived as actual manifestations of the Divine) because of their close association with the transcendent. This is a process known as non-theistic sanctification (Pargament and Mahoney, 2002, 2005). Accordingly, a chaplain patting someone on the shoulder or holding a client's hand has the potential to invoke sacredness because the chaplain, the action, or the content of the conversation may be associated with the transcendent, as understood by the client. Thus, touch itself can become sacralized. Herein lies the benefit of touch as a spiritual coping strategy.

Objects of Spiritual Significance

With this understanding, it becomes clear that the sacred is not limited to human-to-human contact. Through sacralization of the mundane, the scope of the sacred may be expanded to include all aspects of life, including small personal objects (Lamothe, 1998; Pargament and Mahoney, 2002). Whether grand or simple, sacred objects serve as uniting elements between limited, profane existence and the infinite reality of the Transcendent (Lane, 1981). In his book, *Symbols of the Sacred* (2000), Louis Dupré argues that limitations of human intellect require the use of symbols because what is wholly abstract is beyond the grasp of most people. Symbols provide a way of knowing, understanding and interacting with the abstract. They are physical forms signifying an inexpressible vastness. When we are travelling through that dark night of the soul we often need that which is sturdy and solid to hold. Sacralized objects serve that purpose. They are literally something to touch and hold during our difficult hours. They are the mediator between the human being and that inexpressible vastness that some of us call God. Dupré further posits that sacred objects exist in such an intimate relationship with the things that they signify that they are paid the respect and consideration due to the signified. Once again, psychologists Pargament and Mahoney (2002) would agree. They found that extraordinary respect and love are shown to the sacralized. Time and energy are invested in maintaining regular, ongoing contact with these objects when they are used as a coping strategy during times of stress (Miller, 2008). I offer the following case vignette from a patient with whom I worked.

Mattie was an 87-year-old widow with no children. She described herself as a "Cradle Catholic." Regardless of how her faith had ebbed and flowed over her lifetime, Mattie had always participated in the standard practices of Catholic tradition. In middle age she began praying with her rosary beads to lull herself to sleep each night. When her husband died, Mattie found herself overwhelmed with grief to the point where she couldn't concentrate enough to complete all the prayers of the rosary. Yet, she carried her beads with her everywhere. When tears of grief welled up, she would reach into her pocket to touch her rosary. The familiar feel of working the beads between her fingers would calm and comfort her. I met Mattie when she was admitted to hospital to undergo medical investigations for possible cancer. It was then that I heard the story of her rosary. Once again, Mattie found herself so worried that she was unable to pray, despite wanting and needing to do just that. Once again, merely touching her rosary beads brought her comfort.

I grew up in the Catholic Church, so this was not the first time I had witnessed someone being deeply comforted by the touch of a rosary. In fact, this childhood experience is what prompted my interested in this line of research. However, after a few years of working in a large, inner-city hospital, I have come to realize that many objects may take on spiritual significance and are then used as a spiritual coping strategy. In fact, I have worked with patients who are deeply comforted by touching a wide variety of personally sacralized objects. Bonnie, the mother of a patient, is a case in point.

Matthew had come to the hospital for an elective surgical procedure and never regained consciousness. He wasn't expected to live. His mother, Bonnie, and his wife, Jean, took turns sitting at his bedside so that he would not be alone when he died. Bonnie, an active member of the Pentecostal Church, was in the habit of carrying a small stone with her wherever she went. The stone was smooth and shiny, with a deep mossy-green color. It had a lightning bolt of maroon across the front. Bonnie called this a "bloodstone." She would rub the stone against the back of Matthew's hand and across his forearm hour after hour as she sat at his bedside. Bonnie explained to me that doing this was a way of blessing Matthew. "It really works," she said, "I'll show you." She took my hand and rubbed the warm smooth stone across my skin, and I understood what she meant. Bonnie associated her bloodstone with the blood of Christ and in this way sacralized it. She was so comforted by the touch of this stone that she carried it with her at all times. The comfort of its touch was so strong to her that she could not imagine others did not feel it as well.

Touching any object may potentially provide a powerful sense of comfort to those who experience these objects as sacred. Some of these items of comfort are embedded in the narrative of a religious tradition. Others, like Bonnie's bloodstone, have been given a unique meaning by the individual. The overarching feature is that the objects have come to signify a transcendent reality, as perceived by the individual. One of my original research participants stated it simply, "It's not the thing that is sacred so much as what is attributed to it that is sacred" (Miller, 2008, p. 53). In fact, touching a sacralized object during a time of stress and spiritual suffering has been shown to be an effective spiritual coping strategy (Miller, 2008; Miller, Gall, and Corbeil, 2011).

The Function of the Sacralized Object

What function do these objects serve in spiritual coping? Do these objects serve the same purpose with each person? Although there is variation, my colleagues and I found significant commonalities in the experience of holding, touching and/or praying with a sacralized object (Miller, 2008; Miller, Gall and Corbeil, 2011).

Focuses the Mind

One hundred percent of research participants (Miller, 2008) reported that the sacralized objects helped them to concentrate, which was difficult to do in the midst of crisis. It was a little bit easier to block out the worries and concerns that people were struggling with during crisis when they held, touched, and looked at the object. Each of the participants, in their own words, described this. "It centered me and helped me focus on what I wanted and what I hoped for" (Miller, 2008, p. 54).

Tool of Connection

None of the participants perceived the object as a manifestation of a transcendent reality. The objects were not worshipped. Instead, they served as a channel to a transcendent reality. The object was a connecting line to a sacred core. "It helped me to have some kind of connection, especially when I was ticked off beyond belief with this God." Another said, "You feel like you have a line connected to Him – yeah. That's the way it feels. That's my line" (Miller, 2008, p. 57). Whether it's called a "channel," "aid" or "line," all participants used the object in an attempt to communicate and commune with the transcendent reality, as they perceive it. Eighty percent described this as the object's main purpose.

Source of Emotional Support

All participants described reaching for the sacralized object when they needed emotional support. People came to view these objects as a point through which they received an unlimited source of emotional support. This support could be divided into three categories: comfort, strength, and guidance (Miller, 2008).

Comfort. All participants talked about yearning for calm, peace, or comfort when they reached for their object. They described their need to retreat and replenish the spirit through contact with the object: "I don't need Him to do anything practically for me. I'm not starving. I've got a shelter. I'm not lying here bleeding to death. I just need him to calm my fear or calm my sadness" (Miller, 2008, p. 59).

Strength. When faced with challenges, people will sometimes turn to sacralized objects, specifically seeking strength, grounding, or courage so that they can move forward in the face of their stressful life event. They are not asking that the difficulty be lifted from them, or that they be shielded from it. They are asking that their hearts be divinely fortified to ready them for the challenges ahead. Holding the sacralized object before or even during a stressful encounter, combined with prayerful positive self-talk, provided the needed courage. One of the participants in my initial research project described contact with her object as follows: "And it kept me in place, 'You've just got to do this,' so that I didn't get totally freaked out." Another reported: "I feel less powerless. Not that it gives you power, but it certainly gives me strength." A third said: "I think I received strength enough to deal with things and to be—not off in the corner crying all the time" (Miller, 2008, p. 60).

Guidance. People also describe taking hold of the object at a time when they feel overwhelmed by what is happening, keenly aware of their own limitations in that situation and unsure of what to do. A feeling of helplessness may often be part of the journey through illness. A woman who had sacralized a small prayer book given to her by her grandmother said: "Well, usually when I'd turn to that book it was an emotional time in my life where I'm looking for guidance or looking for answers. I'm looking for directions" (Miller, 2008, p. 60).

In summary, sacralized objects serve many functions. They can be an aid in concentration, a tool of connection to a greater reality, as well as an access point to emotional and spiritual support, specifically comfort, strength, or guidance.

A Beneficial but High-Maintenance Coping Strategy

For all the research participants in my first study, the ultimate goal was to access the transcendent for support through physical contact with the sacralized object. That is the coping strategy in a nutshell. The sacralized objects were the mediating pathway between a person and the divine other (Miller, 2008). Since that first study, I have seen hospital patients sacralize many objects by perceiving them to be connected to a wide range of concepts of the transcendent: family, friends, cosmic-universal love. Neither object nor transcendental concept need to be embedded in a religious tradition for the sacralized objects to function as tool for spiritual coping.

When people take hold of these objects it is usually because they are filled with negative emotion and seeking emotional support (Miller, 2008). Research participants, as well as patients I have worked with, have reported that they feel encouraged and their mood is uplifted when they touch these objects. Interestingly, few people have said that touching their object *reduces* their levels of fear or anxiety. More often, this activity seems to introduce positive emotions alongside the negative ones, thus temporarily shifting the balance. For a short while, people feel a sense of comfort, peace, and safety (Miller, 2008). However, this feeling of emotional uplift is short-lived. It comes in small increments. People's narratives indicate that they need to work hard to recapture or maintain this level of comfort. It is a high-maintenance coping strategy. Yet it is so beneficial that people are willing, some even feel compelled, to return to these objects frequently. They keep them close at hand. This is illustrated by the story of Michelle, a research participant (Miller) who struggled with infertility for years before eventually having two successful pregnancies. The following are excerpts from the interview transcript:

> [W]e went in for our appointment. All our tests had come back. Her news for us was that I couldn't get pregnant and he was hopeless too. It was the only thing in my entire life that I didn't have control over. It was just something that – it didn't make sense to me. I was even prepared for her to say, "Well, you know the drugs you're on right now aren't working, but this is what we're going to do next." But, she basically said there was nothing. I just kept staring at this lady and I was baffled that she had no other options for me. I confided in one of the girls at work and she bought this little medallion with a little angel on it that said HOPE. So, I carried that around in my pocket every single day, I never let it out of my pocket. And, anytime I got really, really sad or really stressed, or started to cry, I'd pull out the little angel. It went from the pocket in my shirt to the pocket in my pants, but it never left my body. . . . It's a nice feel to hold on to something. It was nice to have a little piece of comfort. (Interview transcript, Miller, 2008)

Although at one point earlier in the interview Michelle said, "That was me trying to talk with God," communication with God was not the main purpose of her object. It was primarily a source of comfort because it was associated with the love and support she received from her friend. For Michelle, this object signified communion with an empathetic friend. Looking back on that dark time of wanting to have children but being unable to get pregnant, Michelle reflected on the role played by her angel medallion:

> It's something that, when I see them [angel medallions] in stores I think immediately of my children. And I think immediately that something was special about that. I think probably it had a lot to do with the love of the person who gave it to you. Almost like a power that was put into it and passed on. . . . So, I just kind of keep thinking it has a lot to do with—something like that really connects people. But, it's still something tangible. That's all it was. If anything, it's something to hold on to. Because it's easier to say you can hope, but if you actually have something to hold sometimes it makes it a bit easier. (Interview transcript, Miller, 2008)

Michelle needed to touch the angel medallion frequently in order to maintain the level of comfort she needed moment to moment, to make it through each day. She did this by carrying the medallion with her all the time. By her own words, it was always in her pocket. This is a common behaviour with sacralized objects, especially during times of stress (Miller, 2008; Miller, Gall and Grant, 2001).

In summary, contact with sacralized objects provides brief moments of emotional uplift. If the individual is able to string enough of these moments together, the long-term benefits may be significantly propitious. In 1997, Park and Folkman published an investigation of how people cope with severe chronic stress. Although their study didn't investigate the use of sacralized objects in particular as a coping strategy, they reported a similar finding. They monitored the caregiving partners of men with AIDS over a two-year period and found that caregivers who were able to experience frequent, although small, moments of happiness coped better with chronic stress. Park and Folkman theorized that a series of incremental moments of positive psychological states, all strung together, were enough to sustain people through long periods of severe stress. Sacralized objects support these brief moments of positive psychological states.

Evolution of the Sacred Object

The confluence of events that lead to the sacralization of objects is unique for each individual. Family of origin, psychological formation, faith tradition, life experience, and cultural ambiance converge to give significant personal meaning to an object during a period of deep distress and searching (Miller, 2008; Miller, Gall, and Corbeil, 2011). In other words, the objects are usually embedded in the life narrative of the individual and introduced to the individual in childhood by a maternal-like family member, not necessarily the primary caregiver. For one participant, John, a personal relationship developed between John and the object over the years that followed. This relationship ebbed and flowed and matured as the person's faith did. The object eventually came to symbolize the sacred core of transcendent reality, as understood by John. Seventy percent of the participants in my research project reported that the objects were an integral part of their spiritual lives before the stressful event occurred. As children, they may have seen objects in the hands of family members, around the house, in church, and so on, but they did not personally sacralize the object until they experienced a moment of great distress, combined with a yearning for a spiritual connection to the transcendent (Pargament's "relatively available" and "relatively compelling" circumstances). This can be seen in the story of Mattie and her rosary presented earlier.

The narrative of Francine, another patient for whom I provided spiritual care, illustrates a rare exception to this finding. She did not have a long history with the object. It was a gift from me, someone she only knew for two weeks.

Francine was a patient who was going to be transferred to a geriatric rehabilitation centre in a day or two. She was waking up in the night so filled with fear that she had trouble falling back to sleep. Despite being absolutely sure that she had a guardian angel with her throughout the night, she was scared to be alone in the dark. After hearing this story, I brought Francine a small angel carved out of rose quartz. I thought she might want to hold onto it whenever she woke up. Maybe it would make it easier for her to feel the presence of her own guardian angel. Two days later I found Francine very upset because her quartz angel was missing. She suspected that it had gotten tangled in the bed sheets and sent out with the laundry. Francine had even asked her nurse to search around the floor, under the bed, and in her bathroom to try to find it—with no luck. I assured Francine I could get her another one. "Oh no, it's too late." she replied. "I'm being transferred to the other hospital any minute." "Not to worry. I have one in my office. I won't be long." I quickly rushed to fetch the replacement. Five minutes later I returned to find Francine strapped into a transport gurney, tearfully explaining to two ambulance personnel that they could not take her from the hospital yet because she was waiting for her angel!

Even after studying the phenomenon of patient experiences with sacralized objects and having a clear sense of how significant these objects can be, I was taken by surprise by how quickly this process had taken place with Francine and her angel. In just 24 hours she had come to depend on this small statue as a way of connecting to her own guardian angel, and she was comforted by it.

Sacralized Objects in Meaning-Making

There are also long-term benefits which tend to be unexpected by those who use sacralized objects as a coping method during immediate times of distress. These long-term benefits are related to issues of life meaning, purpose, and a new sense of self, which develop after years of reflection on particular stressful events.

Meaning-making is an emotion-focused coping activity in which incongruence between a specific situation and deeply held beliefs (global meaning) is eliminated, either by changing the perceived situational meaning or modifying global meaning (Park and Folkman, 1997). When bad things happen, people are more likely to reappraise the situation to fit their system of beliefs than they are to change or abandon their global meaning (Pargament, 1997). But this process of meaning-making and reappraisal takes time and reflection.

My first study (2008) found that individuals who are in the habit of carrying their objects with them, touching them, and praying with them continue to turn to them for comfort, strength, and guidance even after the stressful event is over. For months and years afterwards people reflect, pray, and ruminate on what happened, all while holding the very object that serves as a connection to transcendent reality. The object encourages a spiritual mind-space within which the meaning-making process takes place. As such, the sacralized object supports a healthy psycho-spiritual framework.

Some of the participants in my research seemed surprised to find that far down the road they were doing better than they expected, and that doing better included a broader aspect of life than that originally impacted by the stressful event. One gentleman referred to this as the "spill-over benefits" of regular prayer with his object (interview transcript, Miller, 2008).

People may not intentionally seek long-term benefits, but they are the blossoms that grow from the coping activities in which people engage in the short term. Effective spiritual coping methods, along with a healthy meaning-making process, may move people toward healing and wholeness. Michelle, the research participant who had trouble getting pregnant, described the final phase of her own journey through the meaning-making process like this:

> I always kind of view this as a life lesson I needed to learn because my life had been too simple up until that point, and I needed to understand that there are some obstacles out there. That obstacle wasn't a real obstacle because it was solved and everything came full circle and was perfectly fine. And other people go through misery, and I just had a tiny little bump in the road, but for me it was a tragedy. It's all in your perspective. I don't know if it changed my relationship with God. I think it made me more mature and appreciate life more. It made me join reality. (Interview transcript, Miller, 2008)

By her own account Michelle was not a religious person. She didn't see her situation as part of a divine plan. The meaning-making process led her to view what happened as an opportunity to join reality. She can now express gratitude for having had that opportunity.

Challenges with the Research

All the participants in the phenomenological investigation of this particular spiritual coping method were Catholic, although their objects were not all embedded in the Catholic tradition. Once I started working in the hospital I encountered people from a wide range of spiritual and religious perspectives who sacralized a wide variety of objects and engaged with them for hoping and coping. However, a controlled study of this wider circle of sacralized objects used by non-Catholics is yet to be done.

Conclusion

Existential questions often arise in times of duress. Beliefs about control, the benevolence and justness of the world, or the existence of a loving God may be thrown into doubt (Gall and Grant, 2005). Whether or not there is a belief in a higher power, and regardless of religious affiliation, when people are troubled by such concerns they yearn for guidance and comfort. They struggle to find meaning in what is happening (Emmons, 1999). Spirituality is one arena within which such quests may take place (Gall and Grant, 2005).

Once a person has entered the spiritual dimension of suffering and reflection, spiritual coping becomes a compelling strategy. An understanding of how touch, whether human-to-human or holding an object, can be sacralized provides the spiritual care provider with a beneficial, therapeutic, spiritual intervention. As with all interventions, a thorough assessment must be done first.

Additionally, sacralized objects can meet a broad set of needs. Once they become established as a component of a person's spiritual practice, sacralized objects prove to be a vital aspect of coping and meaning-making. Although my study found that the benefits of holding and touching a sacralized object parallels those of other spiritual coping methods, this particular strategy does much more than any other single method. Through ongoing contact with a sacralized object, the groundwork is laid for continuous reflection and meaning-making that may lead to positive reappraisal of the event and eventual spiritual growth.

References

Dupré, L. (2000). *Symbols of the sacred.* Grand Rapids: Eerdmans.

Emmons, R. A. (1999). *The psychology of ultimate concerns: Motivation and spirituality in personality.* New York: Guilford Press.

Gall, T. L. and Grant, K. (2005). Spiritual disposition and understanding illness. *Pastoral Psychology, 53*(6), 515–533.

Hodge, D. R. (2003). *Spiritual assessment: A handbook for helping professionals.* Botsford, CT: NAACSW.

Hood, R. W., Spilka, B., Hunsberger, B. and Gorsuch, R. (1996). *The psychology of religion.* New York: The Guilford Press.

Jackson, L. E. and Coursey, R. D. (1988). The relationship of God control and internal locus of control to intrinsic religious motivation, coping and purpose in life. *Journal for the Scientific Study of Religion, 27*(3), 399–410.

LaMothe, R. (1998). Sacred objects as vital objects: Transitional objects reconsidered. *Journal of Psychology and Theology, 26,* 159–167.

Lane, D. A. (1981). *The experience of God: An invitation to do theology.* New York: Paulist Press.

Levin, J. S. (2001). *God, faith and health: Exploring the spirituality health connection.* New York: Whiley and Sons.

Miller, L. M. (2008). *Sacred objects as tools of connection with the transcendent: The essential structure of prayer with a sacred object.* Master of Arts Thesis, Saint Paul University, Ottawa.

Miller, L. M., Gall, T. L. and Corbeil, L. (2011). The experience of prayer with a sacred object within the context of significant life stress. *Journal of Spirituality in Mental Health, 13*(4) 247–271.

Pargament, K. I. (1997). *The psychology of religion and coping: Theory, research, practice.* New York: Guilford Press.

Pargament, K. I. (2007). *Spiritually integrated psychotherapy: Understanding and addressing the sacred.* New York: Guilford Press.

Pargament, K. I., Koenig, H. G. and Perez, L. M. (2000). The many methods of religious coping: Developing and initial validation of the RCOPE. *Journal of Clinical Psychology, 56*(4) 519–543.

Pargament, K. I. and Mahoney, A. (2002). Spirituality: Discovering and conserving the sacred. In C. R. Snyder and S. J. Lopez (Eds.), *Handbook of positive psychology* (pp. 646–659). Washington D.C.: APA Press.

Pargament, K. I. and Mahoney, A. (2005). Sacred matters: Sanctification as a vital topic for the psychology of religion. *The International Journal for the Psychology of Religion, 15*(3), 179–198.

Pargament K. I. and Park C. L. (1995). Merely a defense? The variety of religious means and ends. *Journal of Social Issues, 51,* 13–32.

Park, C. L. and Folkman, S. (1997). Meaning in the context of stress and coping. *Review of General Psychology, 1*(2), 115–144.

Samson, A. and Zerter, B. (2004). The experience of spirituality in the psycho-social adaptation of cancer survivors. *Journal of Pastoral Care and Counseling, 57*(03), 329–343.

Tix, A. P. and Frazier, P. A. (1998). The use of religious coping during stressful life events: Main effects, moderation, and mediation. *Journal of Consulting and Clinical Psychology, 66,* 411–422.

Chapter X

Touch(ed) in Palliative Care Nursing: Moving with/in An Uncertain Practice

Lacie White and Christine McPherson

> When you begin to touch your heart or let your heart be touched, you begin to discover that it's bottomless, that it doesn't have any resolution, that this heart is huge, vast, and limitless. You begin to discover how much warmth and gentleness is there, as well as how much space. (Pema Chödrön, 1994, p. 128)

In a poignant video released by the Canadian Virtual Hospice (n.d.), a website of resources for palliative care patients, families, and their providers, Jim Mulchay shares his experience of a "simple touch" offered by his nurse as he underwent an invasive procedure. The nurse held his hand "in a fashion that was not entirely clinical in nature." Jim recounts how their encounter continued to unfold after the procedure was complete:

> Nurse: [Uttering a sigh] You know, I hate that procedure.
> Jim: Thank you. I don't know if they train you to do this or not, but if they don't train you they should. Because of how human it made me feel. How cared for it made me feel . . . . [You] are, in fact, touching me gently, in a caring fashion.

> Nurse: [Appreciating Jim's comment] And thank you for holding my hand because it was difficult for me to watch this, and to witness you in that process.

Within this nurse-patient interaction the physical act of touch manifests in a profoundly intimate way, *touching* both the nurse and patient affectively. Jim recalls being moved by the nurse's presence and way of being in relationship with him. "As opposed to being an object," he experienced being "a subject . . . a person and not a pathology." The nurse, in this instance, shows an openness to Jim's suffering, and in this way is *touched* in the midst of touching another. Thus, underlying what is a seemingly simple behaviour comprising physical contact between the nurse and patient is a highly complex, intersubjective, and context-dependent act that has meaning for those involved (Chinn and Kramer, 2011; Green, 2013).

Indeed, nursing by its nature is a deeply intimate profession where touch is integral in providing care to another. In their day-to-day interactions, nurses touch patients frequently, as they tend to their needs, conduct assessments, administer interventions, and provide comfort. Yet, by its routinized nature, touch can become an automatic, unconscious, and disembodied act. Further, as most research on touch focuses on outcomes rather than processes (Estabrooks and Morse, 1992), there is little understanding of how to foster and use touch in nursing practice, or to understand the meanings associated with the practice from the perspectives of patients and nurses.

In this chapter, we explore touch as an aesthetic, embodied, and unclear act in nurse-patient interactions that can foster connection and provide comfort. To ground our discussion, we use our experience within palliative care, which is "an approach that improves the quality of life of patients and their families facing the problems associated with life-threatening illness, through the prevention and relief of suffering" (World Health Organization, 2015, para 1). To begin, we position touch as an embodied and aesthetic practice in nursing care. Approaches to touch that can foster or challenge enacting touch in this way will then be discussed. Finally, mindfulness and storytelling will be suggested as methods that can support cultivating approaches to touch that are aesthetic and embodied. To explore theoretical perspectives we offer reference to two stories: "With Mary, With Jerry," and "With Jake and Family" (see appendices A, B, and C, respectively). These stories were written over the course of five years while the first author practiced as a palliative care nurse in various settings and practiced mindfulness alongside caring for people who were dying and their families. Before reading further, readers are encouraged to explore their own responses to the stories presented in the appendices, perhaps taking time to write freely about what arises as they read the stories, and to reflect on their own experiences and thoughts. This self-reflection will offer a point of reference for discussion later in the text.

Touch in Nursing: An Aesthetic and Embodied Practice

There is widespread recognition in the nursing profession of the importance of touch in supporting patients physically, socially, emotionally, and spiritually through health and illness (Chang, 2001; Connor and Howett, 2009). Touch is linked to core nursing concepts, including the therapeutic relationship, as nurses can express their presence and availability to patients and families through the touch they provide (Benner, 2004; Fredriksson, 1999). Despite the significance of touch in nursing, its use in practice remains, at best, uncertain, for the very reason that it is so contextually bound. Van Dongen and Elema (2010) suggest "touching as an art means that people have to redefine, reinvent and reshape their ways of touching others day by day, situation by situation" (p. 153). For these reasons, touch in nursing is often aligned with aesthetic knowing, or the art of nursing, where nurses perceive and respond to individual and contingent needs of the unfolding, moment-to-moment experiences they find themselves with/in (Chinn and Kramer, 2011).

Associated with the aesthetics of nursing is the concept of embodiment. As a way of being present and living with/in and through the body, nurses use their bodies as instruments to perceive needs and care for their patients (Draper, 2014). Through attending to and touching patients and families, nurses can also be *touched* by the experience. For nurses, touching the body of someone they are caring for can elicit or amplify an emotional response, which at times can be uncomfortable, and lead nurses to limit physical contact made with patients in their practice (Picco, Santoro, and Garrino, 2010; Van Dogen and Elema, 2010). Therefore, although touch is often approached as an act of physical contact, or one body touching another, emotional responses also influence the way touch is enacted in practice; it is helpful to extend this conceptualization beyond the physical realm, into the symbolic (i.e., to be touched affectively; Chinn and Kramer, 2011).

As an embodied and relational concept, integrating touch into practice is complex and cannot be done prescriptively. Further, it is not our intention to "naturalize the body" (Violi, 2009) which is a risk within this chapter, and a position that can be seen in research and theoretical discourses, subsequently affecting the way touch is approached and taught in nursing. Violi describes naturalization of the body as presenting it as "a non-gendered, pre-discursive phenomenon, hiding the concrete reality of the many different bodies all persons possess, with all their social, cultural and discursive determinations" (p. 58). As varied humans, encountering equally diverse individuals in our practices, it is important to consider our own (and others') social and cultural locations and how this affects our embodiment and (dis)comfort to touch in and through these bodies.

(In)visible and (Dis)embodied Practices of Touch in Nursing

Within the nursing literature, touch has been categorized in various ways based on its function and intentionality (Connor and Howett, 2009). Typically, it is dichotomized into touch as part of performing a task-related function or as a means to comfort. Dichotomous perspectives regarding touch and its use in nursing practice can fragment the body and further affect how it is integrated into practice. In general, the emphasis on doing something to/for the body through attending to various tasks is often positioned and carried out in an objectified way that overshadows touch that is intentional in its possibilities to provide comfort and attend to the emotional needs of patients (Connor and Howett, 2009; Wolf, 2014). As a result of tasks and technologies being ever more visible and the focus in nursing practice and care plans, comforting touch has become an endangered art (Benner, 2004).

Draper (2014) expresses concern that "the knowledge we have about what it is to live 'with' and 'in' a body including emotions and relationships with others can become marginalized and misplaced" (p. 2238) by scientific and rational approaches to care. Failure to appreciate the intersubjective experience of touching and being touched by others can challenge embodied approaches to enacting touch in practice. Undeniably, nursing's relationship to the body has been largely influenced, as Wolf (2014) describes, by a biomedical approach to health:

> The growth of medical technologies and the rapid transfer of tasks such as physical assessment and interventions are seductive tools that both add to the nurses' understanding of the objective nature of bodies but also deflect nurses from their attention to the subjective experience of their patients. (p. 151)

Increasing interventions, the medicalization of care, and focusing on "doing for" as opposed to "being with" are increasingly prevalent in palliative care, which has historically been grounded in a slower pace from which to attend to the physical, social, emotional and spiritual needs of patients and families (Bruce and Boston, 2008). This is a significant concern as the touch of hurried hands, devoid of empathy, can perpetuate feelings of isolation and loneliness (Sand and Strang, 2006).

In contrast, touch offered with an intention to foster presence, trust, and communication in the nurse-patient exchange can be deeply supportive, comforting, and healing (Chang, 2001; Connor and Howett, 2009; Fredriksson, 1999). As the stories presented in this chapter show, intentional touch can offer space and safety to go to places that scare another; as one knowingly walks closer to death, questions like "what will happen when I die?" can and often do surface. In the story "With Mary," it was the presence and connection communicated through touch that opened possibilities to explore such unknown territory. Grief is a significant and profound emotional response to sickness, death, and dying, which can be held in a space of acceptance, and sometimes this holding includes physical contact with another (see the story "With Jake and Family"). It can be argued in the story "With Jerry," who began to rest in the act of being held by another (and before medications as a technological advancement in care could be offered), that he was simply exhausted through his experience; however, another theory upheld within nursing is that skillful, intentional presence can be healing. Touch was one way to foster presence and relational intimacy with Jerry. It required a particular way of being present: to hold him, to know and connect to his experience, including entering into, to some degree, the intense fear that he was communicating. Also, it required the health care team to attend to their own bodies and abjection at the sight of finding Jerry's flailing body covered in his own secretions. Thus, when the nursing body is able to remain open, aware (of self and other) and accepting as it listens to and holds, physically or symbolically, the body of someone in an experience of fear, grief, and existential distress, this can support another to a more calm and comfortable place.

At the Margins of Care: Personal and Professional Boundaries Affecting Touch

How comforting touch is integrated into practice can be affected by personal and professional boundaries. Drawing on the work of Parker, Wolf (2014) describes how nurses are located within "overlapping and interpenetrating margins between professional constraint and personal intimacy" (p. 150). Despite scientific and rationalized approaches to care, which suggest that professionally it is important to maintain some objectivity and distance from patients and family being cared for, to offer comfort touch and care in practice requires relational intimacy and emotional closeness. Consider the moments of touch in the stories with Jerry, Jake's family, and Mary. Is the level of contact and relational intimacy appropriate by professional standards? How can this tenuous margin of care be explored and embodied skillfully? Gleeson and Timmins (2005) conducted a review of clinical effectiveness of touch as a nursing intervention and concluded that, based on a lack of clear guidelines and perceptions of nurses' and patients' experiences of touch in practice, "ad hoc use of comfort touch as an intervention on its own right, must be resisted" (p. 76). However, given the reality that nursing fundamentally requires that we touch, and that individuals (both nurses and patient/families) vary in their level of comfort and desire for touch, it may not be possible to establish clear guidelines around the ways in which to touch in nursing practice.

Within palliative care nursing practice, comforting touch requires a particular level of engagement and intention to be intimate with others in what is often emotional and suffering experiences. However, as previously discussed, attending to and sharing such experiences with others can generate discomfort and (un)consciously guide nurses to limit their level of engagement with patients and families (Picco et al., 2010; Van Dongen and Elema, 2010). Nurses may reduce their use of touch when providing care (for example, consider the strong personal boundary imposed in the story "With Mary"); or task-based approaches that require less emotional proximity (Pedrazza, Minuzzo, Berlanda, and Trifiletti, 2014) can become the default mode for how touch is enacted in practice (Picco et al., 2010). Picco et al.

(2010) emphasize that "being intimate and close to a body other than one's own requires overcoming personal physical and emotional barriers" (p. 44). Therefore, a willingness by nurses to be touched by the fullness of experiences, which includes strong emotional exchanges, is needed in therapeutic relational practice. Honest self-reflection of the nurse's ability to attend to these moments and a growing awareness of how touch may support (or not) a particular encounter with a patient and family is needed. Additionally, exploring how nurses may cultivate a growing capacity for the emotional response to exist (within themselves and others) may support skillfulness in their movements in and out of touch in their practice.

To Touch and Be Touched: Cultivating Capacity

We now turn to mindfulness and storytelling as two suggested methods offering aesthetic and embodied ways of being and, thereby, the ability to move into—and out of—moments of touching and being touched in nursing practice.

Mindfulness

Jon Kabat-Zinn (1990) is often credited with introducing mindfulness, rooted in 2,600-year-old Buddhist philosophy, into mainstream health care as a secular and groundbreaking approach to healing and wellness. Mindfulness is a way of being and becoming present, where one is able to attend to the unfolding field of experience, with awareness and acceptance, allowing moments to rise and fall away without becoming entangled or distracted (Amaro, 2015; Kabat-Zinn, 1990, 2003). Among the various approaches associated with cultivating mindfulness, the practices of meditation, breathing, and body scans are frequently referenced (e.g., Halifax, 2008; Kabat-Zinn, 1990; Rinpoche, 2002, as seminal texts to develop experientially these practices). In a qualitative review of health care professionals' experiences of engaging in mindfulness programs, Morgan, Simpson and Smith (2014) found that a number of participants, including a small subset of nurses, experienced a changing relationship to moments encountered in the midst of caring for themselves and others; specifically, participants were able to witness or stay with/in unfolding experiences, including those in which strong emotions were present. Thus, through mindfulness a growing capacity to be with experiences as they arise can be fostered. In this way, a relational space opens up to be touched, and subsequently to touch intentionally and skillfully in nursing practice.

Attuning Mind, Body, and Emotions

Mindfulness can support one to cultivate and experience in an integrated and dynamic way attunement between mind, body and emotions (Kabat-Zinn, 1990). Conceptualizing touch in a way that extends into affective realms, this synchronization of mind, body, and emotions is an invaluable method to support enacting touch in more embodied ways. Returning to the stories in this chapter, awareness of mind, body, and emotional aspects of experience are apparent and foster the ability to physically move into contact with another. Awareness of the affective response is significant in that touching patients and family members required being in contact with, or touched by existential anxiety, abject terror, grief, and perhaps disgust. Also, the physical body is implicated in these experiences; drawing awareness to it helps in such a way that one can flow within the unfolding moments, while simultaneously providing a way to stay grounded in these experiences (e.g., see "With Jake and Family": "I . . . also tried to stay as best I could with my own breath; aware of my hand . . ."). In the story "With Jerry," there was a moment of being aware of his suffering in such a way that the mind danced around an invisible line or boundary. What presented itself was the choice to connect or disengage (recoil) from Jerry and his experience. However, through self-awareness and observation of feelings, thoughts, and physical sensations that surface as fear manifests, one can more easily see the experience for what it is, and more readily choose how to respond in the moment(s) (Kabat-Zinn, 1990). This may include choosing to move toward the suffering of another and touching complex and uncomfortable moments with someone at the end of life.

Moving with Uncertainty and Vulnerability: Flowing of Frozen Waters?

Boston, Towers and Barnard (2001) emphasize the importance of nurses embracing personal risk and vulnerability when caring for patients at the end of life. They suggest that being vulnerable "may mean moving into uncharted waters of experience" (p. 252). Mindfulness may be one way to foster this way of being with risk and vulnerability. For example, in a unique hospice setting that integrates mindfulness practice with a palliative care approach, Bruce and Davies (2005) found that participants expressed a willingness to forgo a predetermined script or guide in order to relationally be with/in unknown moments of caring and to respond to the changing needs of the dying person and their family. Further, participants in this hospice setting explained that part of their mindfulness practice included fostering the ability to see when one turned away from a situation or person, return to the moment(s), and through this approach "be willing to shift and change in the process" (Bruce and Davies, 2005, p. 1339). Thus, part of their mindfulness process included observing and allowing their own humanity, opening to the (un)knowing, and creating a space to gently work with the boundary of their capacity to be present with others. This process lends itself to "a way of being, a way of seeing, which is embodied, inhabited, and [*sic*] grown into" (Kabat-Zinn, 2003, p. 148) as one compassionately explores the (un)comfortable edges of being present to ourselves and others.

Through mindfulness practices, one can remain disentangled from experiences so as not to become overwhelmed or "swept away" (Kabat-Zinn, 1990, p. 334) and instead flow with the changing sea of moments in unfolding experience. Drawing on aesthetic knowing, Chinn and Kramer (2011) describe how the movements of nurses "express a multitude of meanings on many levels" (p. 142) and "make both physical and symbolic touch possible" (p. 143). Consider again the story of being "With Jerry." It begins with a description of standing frozen. As individuals and as a team of health care providers, there may not have been the space to (however, momentarily) move with/in the moments with Jerry. Thus, a valuable reflection for nurses may be to explore when movement or flow is not apparent, or is difficult to maintain in their experience. In the following section we draw on storytelling as one way to explore such experiences; however, mindfulness practices also offer nurses a method of being and becoming aware and present to unfolding moments with their patients. Through mindfulness one can grow the capacity to enter into experiences with another, which can include moving into physical contact, or skillfully perceiving the moment(s) when touch would not be warranted, helpful, or perhaps, even harmful.

Storytelling

When difficult experiences interrupt the movement of one's life, stories can help to "restore the flow" (Frank, 2009, p. 163). In exploring the value of stories we draw on the work of sociologist Arthur Frank (2009, 2010). In his text *Letting Stories Breathe*, Frank (2010) encourages a relationship with stories that support their inherent movement as living, breathing entities. Stories have the capacity to move and act in a variety of ways. They offer a way of exploring and attending to embodied responses and thereby foster openings to flow more freely into and out of physical and symbolic contact with others in nursing practice. They can "*teach* on an affective and even corporeal level" (Frank, 2009, p. 172) in a way that practice-based guidelines cannot. Further, stories help us to understand and make meaning of significant experiences and the complex and mysterious nature of being embodied (Frank, 2009). Given their capacity to evoke, stories are explored here as a method to touch symbolically both the teller and listener/reader.

Going Slow to Know: Storying Meaning

Why did this story "With Jerry" come to mind "over and over again?" In a way, it was a story that haunted and came to mind repeatedly as an experience that sought understanding. In fact, none of the stories offered here were written for this chapter. They are from a library of stories written throughout a time the first author was working as a palliative care nurse; writing became part of the practice to understand, make meaning, and process the highly relational and dynamic work of providing care to dying patients and their families. Thus, stories are a valuable tool to slow down experiences that often unfold too quickly to understand or integrate while they are encountered in practice (Frank, 2009; Stanley and Hurst, 2011).

Evoking the Body: Storying Practices of Touch

Narratives are "essential to convey and preserve knowledge about the skill of involvement (getting the right kind of involvement and interpersonal distance to fit the situation), because relational skills always involve the concrete other and are always context dependent" (Benner, 1991, p. 9). However, one contextual piece often overlooked in nursing is the body (Draper, 2014). Within nursing literature, "there is almost the 'shadow' of nurses' bodies and embodiment waiting to be discovered" (Shakespeare, 2003, p. 48). Stories are one way to address this concern, as they are an evocative force that can penetrate and inhabit the body. Therefore, for embodied experiences, which extend beyond what can be described with words, stories are tools (methods) to teach, research, and learn about touch in nursing practice.

Cultivating a Community of Practice: Storying Relations

When nurses share their stories of practice they can generate dialogue and cultivate communities of practice. It is important to consider how stories can "touch" individuals and communities in ways that can be transformative and/or dangerous (Frank, 2009). One way to foster aesthetic knowing is to support and guide nurses and students in (re)imagining, with the use of stories, what they might do in complex and dynamic moments encountered in their nursing practice (Chinn and Kramer, 2011). Stories such as those used in this chapter, however, are remarkably absent in academic texts within nursing (Holmes, Perron, and O'Byrne, 2006), or communicated in practice settings (Bruce and Boston, 2008). Therefore, although individual introspection and writing practices are valuable, sharing stories may offer a valuable way to communicate and reflect on the use of touch within nursing practice.

Yet a story can also destabilize an unsuspecting reader or listener who may be unprepared to encounter the content embedded within it (Frank, 2009). Returning to the stories within the chapter, we ask the reader to reflect again on their experiences of encountering them. Who the stories touch, and in what ways, will vary as widely as the number of readers of this text. The stories presented here may not resonate for all readers as they are set in the context of end-of-life experiences, where strong emotions, suffering bodies, and existential questions and unease can manifest, and potentially elicit strong embodied reactions. Nonetheless, for nurses and students who will inevitably encounter situations in which they will be asked to be present with and through such moments, stories such as those presented here (and the ones that arise naturally in the mind of the reader) are important to explore. Also, the stories embedded within this text are not meant to offer the way to touch, or the true story to reenact in palliative care nursing practice. Frank (2009) cautions, "those who affiliate around some particular story can become insular in asserting this story as the privileged template for the treatment and care of the dying person" (p. 171). With this in mind, it is important to offer opportunity for the reader/listener to engage with stories in their own way, inviting individual and varied responses.

If nurses and nursing students are unable to be with these inbodied (and often intense) responses, how then will there be space for such responses to exist within clinical settings, where they will be experienced in an even more direct and embodied way? We suggest engaging in the storytelling process slowly, so that the tellers' and listeners' responses can be explored. Further, the movement of stories can be fostered by not finalizing an individual's story, which can change, finding new forms and meanings over time (Frank, 2009, 2010). Also, within the practice of writing and sharing, there is an inherent vulnerability as the emotional response can be powerful in retelling and listening to stories (Stanley and Hurst, 2011). Therefore, courage, vulnerability, and as stated in mindfulness circles, a willingness to lean (*gently*) into the experience are needed. Thus, mindfulness practice can also be integrated into such explorations with stories and storytelling. For example, breathing practices, body scans, and brief meditations can be introduced as ways to see and be with whatever is arising and serve as ways to be with emerging experience. Through the practice and process of storytelling, nurses can grow in their ability to be compassionate and empathize with others, connecting and sharing in experience (Stanley and Hurst, 2011); in this way aesthetic and embodied knowing can be cultivated and guide nurses into and out of touch skillfully.

Conclusion

Nurses can learn to touch and be touched existentially, emotionally, and physically through time spent on skills of embodiment. Mindfulness and storytelling are two such practices that can foster aesthetic and embodied knowing. Through exploration and consideration of our own boundaries and (dis)comforts with touching in nursing practice, nurses and students can develop an ever-enlarging capacity to use touch as one way to skillfully provide comfort and support to people who are dying and their families. To be touched by moments, by our own embodied responses, and by the people we care for is to discover, as Pema Chödrön describes in the epigraph of this chapter, a heart that is "huge, vast, and limitless."

Appendix A: With Mary

Mary was the second patient I had ever cared for as a home hospice nurse. In my visits to her home, we would sometimes talk about her cancer and ways to manage her symptoms. At other times we talked about her life, or I would share snapshots of mine. However, after knowing one another for some time, her behaviour began to change. She became silent and closed. No longer feeling our relationship was therapeutic, I asked the team leader, Jen, to make a visit and see if she could uncover what was happening.

Immediately after her visit, Jen called me and relayed, "Mary wants to see you more. And she would like a hug from you." Mary's request caused me to reflect deeply on my approach to life and work. I was holding a piece of myself back from Mary. I knew she was dying and to allow myself to get closer, I thought, would open me to great pain and grief as I supported and witnessed her inevitable death. I was challenged to see and shift from my stance of self-protection and enter more deeply into relationship with Mary.

And so I opened. And Mary opened. We learned together how to do this. On the next visit to Mary's home, we sat on her couch and I shared how grateful I was that she was able to ask for what she wanted. Mary scotched herself closer to me, and I put my arm around her shoulders in a gentle embrace. We sat silently for a few moments, and Mary's head came to rest upon my shoulder. The moments for me felt both comfortable and uncomfortable, as I was unsure of what was to arise in this shared connection. Then, Mary looked at me with soft watery eyes and placed her hand on my leg as she asked, "Dear, what is it going to be like to die, what do you think happens?"

I responded, "I have some theories but I don't know."

"Do you think people come back again?"

"It may be possible," I said, reinforcing that I was not sure.

Becoming tense, sitting up straight, and stopping her gentle strokes on my leg, Mary made an endearing, scrunched up face and shared quite genuinely, "if it is true that they can, I do not want to come back as a snake."

After a few more moments of silence and her body resting back next to mine, I asked, "What would you want to come back as, Mary?"

"A teacher, I should have been a teacher in this life." Her answer had been immediate.

Appendix B: With Jerry

An image that continues to come to my mind over and over again is of a man, Jerry, who after two days of unresponsiveness, bedbound, and appearing to be imminently dying, found his way out of bed and onto the floor. My colleague Sandra found him and called for me to quickly come and help. Jerry was a frail, tiny man—skin and bones. He was rolling around on the cold floor in his bodily fluids (oral secretions and urine). He moaned. I think we all stood frozen for a moment. I asked Sandra to prepare all we can: dilaudid, scopolamine, midazolam. To the volunteer I asked for warm blankets and a pillow. It was a sight I/we had not been prepared to see. After a few moments of back and forth in my mind, from connecting with Jerry and recoiling at the sight of his suffering, I crouched down, placing the pillow under his head. I have etched in my memory the fear and relentless terror in his eyes. I take a breath now, as I most certainly did then, to be able to stay with him. Staying with him, I held his body with mine, breathing deeply, and relaxing my body, even as I could feel the tension in his, until my colleagues were ready to assist him back to bed. This time, which was only seconds to minutes, seems etched in my mind like it was shared for days. Deeply meeting what he communicated in his movements, moans, and most of all eyes, I said, "I am here, we are here, we are not leaving." And I am on the floor with him, holding his head, not leaving his eyes, waiting for colleagues. And I can feel and see him settle. We had not yet given him "all we can." Yet, in sharing the experience with him, his body softened, his eyes lost the magnitude of terror. I have regretted that we continued without discussing the experience, which had made an impression on all of us. That day we were so busy. We had to keep moving; there were others who needed support.

Appendix C: With Jake and Family

I had not yet met Jake and his family. Walking into the room, I stood a few steps back from the end of the bed. Jake's wife sat at the head of the bed as he lay imminently dying. Their daughter stood silently off to the side. I did not say anything, but took a breath and entered into the experience, making eye contact only briefly with the daughter. After a few moments of watching Jake's irregular and infrequent breaths, his wife turned to me and asked tearfully in one of the long pauses in Jake's breathing, "Is he gone?"

I could not be certain. "No, I don't think so," I said softly, "but these are his last breaths." There was no further discussion between us as she turned back to her friend and husband of 60 years and said through sobs, "I love you," and rested her head next to his. After only a few more inhales and exhales, Jake's breathing ceased. I continued to stand at the back of the room, feeling into and considering what may be next for me to do, even as I was uncertain. Holding her husband, Jake's wife continued to cry. Their daughter tried to console her mother by placing a hand on her shoulder. However, seeing that their daughter also needed support, I asked if I could put my hand on her back. She nodded in a way that seemed to welcome my presence, and so I placed my hand gently to rest between her shoulder blades as their daughter began to cry.

We shared a series of moments: the body of a man after his final exhalation, his grieving wife and daughter, and myself. The enormity of loss was palpable, but I tried to stay as best I could with my own breath, aware of my hand that was in contact with their daughter. After a few minutes their daughter turned to me and said, "Mom needs your hand more than I do." Taking her cue, I knelt down low; not feeling her mother's need for my contact, I slipped my hand to partially rest under her hand.

"Is there anything I can do for you now?" I asked.

Jake's wife shook her head no. She raised her arm up to gently place her hand on to my face as we made eye contact. "Thank you," she said.

References

Amaro, A. (2015). A holistic mindfulness. *Mindfulness, 6*(1), 63–73.

Benner, P. (1991). The role of experience, narrative, and community in skilled ethical comportment. *Advances in Nursing Science, 14*(2), 1–21.

Benner, P. (2004). Relational ethics of comfort, touch, and solace: Endangered arts? *American Journal of Critical Care, 13*, 346–349.

Boston, P., Towers, A. and Barnard, D. (2001). Embracing vulnerability: Risk and empathy in palliative care. *Journal of Palliative Care, 17*, 248–253.

Bruce, A. and Boston, P. (2008). The changing landscape of palliative care: Emotional challenges for hospice palliative care professionals. *Journal of Hospice & Palliative Nursing, 10*, 49–55.

Bruce, A. and Davis, B. (2005). Mindfulness in hospice care: Practicing meditation-in-action. *Qualitative Health Research, 15*, 1329–1344.

Canadian Virtual Hospice (n.d.). *A story about care.* Retrieved from http://www.virtualhospice.ca/en_US/Main+Site+Navigation/Home/Support/Support/The+Gallery/Stories/The+Story+About+Care.aspx

Chang, S. (2001). The conceptual structure of physical touch in caring. *Journal of Advanced Nursing, 33,* 820–827.

Chinn, P. L. and Kramer, M. K. (2011). *Integrated theory and knowledge development in nursing* (8th ed.). St. Louis, MO: Mosby/Elsevier.

Chödrön, Pema. (1994). *Start where you are: A guide to compassionate living.* Boston: Shambhala Publications.

Connor, A. and Howett, M. (2009). A conceptual model of intentional comfort touch. *Journal of Holistic Nursing, 27,* 127–135.

Draper, J. (2014). Embodied practice: Rediscovering the 'heart' of nursing. *Journal of Advanced Nursing, 70,* 2235–2244.

Estabrooks, C. and Morse, J. (1992). Toward a theory of touch: The touching process and acquiring a touching style. *Journal of Advanced Nursing, 17,* 448–456.

Frank, A. (2010). *A socio-narratology: Letting stories breathe.* Chicago: University Chicago Press.

Frank, A. (2009). The necessity and dangers of illness narrative, especially at the end of life. In Y. Gunaratnam and D. Oliviere (Eds.), *Narratives and stories in health care: Illness, dying and bereavement* (pp. 161–175). Oxford: Oxford University Press.

Fredriksson, L. (1999). Modes of relating in a caring conversation: A research synthesis on presence, touch and listening. *Journal of Advanced Nursing, 30,* 1167–1176.

Green, C. (2013). Philosophic reflections on the meaning of touch in nurse–patient interactions. *Nursing Philosophy, 14,* 242–253.

Halifax, J. (2008). *Being with dying: Cultivating compassion and fearlessness in the presence of death.* Boston, MA: Shambhala Publications, Inc.

Holmes, D., Perron, A. and O'Byrne, P. (2006). Understanding disgust in nursing: Abjection, self, and the other. *Research and Theory for Nursing Practice, 20,* 305–316.

Kabat-Zinn, J. (1990). *Full Catastrophe Living.* New York: Bantam Dell.

Kabat-Zinn, J. (2003). Mindfulness-based interventions in context: Past, present, and future. *Clinical Psychology: Science and Practice, 10,* 144–156.

Morgan, P., Simpson, J. and Smith, A. (2014). Health care workers' experiences of mindfulness training: A qualitative review. *Mindfulness,* 1–15.

Pedrazza, M., Minuzzo, S., Berlanda, S. and Trifiletti, E. (2014). Nurses' comfort with touch and workplace well-being. *Western Journal of Nursing Research, 37,* 781–798.

Picco, E., Santoro, R. and Garrino, L. (2010). Dealing with the patient's body in nursing: Nurses' ambiguous experience in clinical practice. *Nursing Inquiry, 17*(1), 39–46.

Rinpoche, S. (2002). *The Tibetan book of living and dying.* New York: HarperCollins.

Sand, L. and Strang, P. (2006). Existential loneliness in a palliative home care setting. *Journal of Palliative Medicine,* 9, 1376-1387.

Shakespeare, P. (2003). Nurses' bodywork: Is there a body of work? *Nursing Inquiry*, 10(1), 47-56.

Stanley, P. and Hurst, M. (2011). Narrative palliative care: A method for building empathy. *Journal of Social Work in End-of-life & Palliative Care, 7*(1), 39–55.

Van Dongen, E. and Elema, R. (2010). The art of touching: The culture of 'body work' in nursing. *Anthropology and Medicine, 8*, 149–162.

Violi, P. (2009). How our bodies become us: Embodiment, semiosis and intersubjectivity. *Cognitive Semiotics, 4*(1), 57–75.

World Health Organization (2015). WHO definition of palliative care. Retrieved from http://www.who.int/cancer/palliative/definition/en/

Chapter XI

Touch in Supervision

Cynthia Bilodeau

Supervision is an important part of counselling and psychotherapy, serving as the primary means for developing professional competency and for monitoring the quality of services provided (Bernard and Goodyear, 2013). The use of touch, which is viewed as having the potential for both positive and detrimental effects on clients and therapeutic processes (Bonitz, 2008; Calmes, Piazza and Laux, 2013; Eyckmans, 2009; Horton, Clance, Sterk-Elifson and Emshoff, 1995; Hunter and Struve, 1998; Phelan, 2009; Stockwell and Dye, 1980), requires supervisors to carefully navigate critical boundary issues while guiding and developing ethical awareness and competence in their supervisees. The ability of supervisors to successfully navigate this process can have serious implications for supervision outcomes, client wellness, and supervisee development. Despite these important implications, however, the literature on touch in supervision has been negligible, leaving supervisors ill-equipped with regards to supervision of client-supervisee touch, and leaving them uncertain with regards to appropriate touch between themselves and their supervisees (Burkholder, Toth, Feisthamel, and Britton, 2010). This chapter considers theoretical, clinical, and ethical perspectives of professional supervision and therapeutic use of touch to assist supervisors in addressing these issues in supervision ethically and responsibly. Furthermore, this chapter begins the long-avoided conversation on touch between supervisors and supervisees, outlining some preliminary guidelines on best practices for supervisor/supervisee touch.

Regardless of the helping profession, clinical supervision has most likely played an important role in one's training process. Recognized as a distinct intervention in itself, supervision serves primarily to enhance competence and to monitor the quality of services provided (Bernard and Goodyear, 2014; Falendar and Shafranske, 2007). In the context of counselling and psychotherapy, supervisors are ethically responsible for developing supervisee technical competence, ethical awareness, and self-growth, all the while ensuring client safety and gatekeeping their profession (American Counselling Association [ACA]; American Association for Marriage and Family therapy [AAMFT], Canadian Association for Spiritual Care [CASC], Canadian Counselling and Psychotherapy Association [CCPA], and Canadian Psychological Association [CPA]).

Although supervision can take on different forms (such as group supervision or individual supervision), it is generally provided by a more experienced member of the profession who is esteemed as having enough knowledge and expertise in the field to appropriately guide supervisees in bridging their theoretical and practical knowledge (Bernard and Goodyear; 2014; Barnett, Cornish, Goodyear, and Litchenberg, 2007). In the field of counselling and psychotherapy, where theories collide, where views on many topics diverge, and where empirical evidence is commonly found to conflict, it is not uncommon for supervision to extend throughout one's professional career as part of a preoccupation with maintaining a hygienic and ethical practice (Barnett, Cornish, Goodyear, and Litchenberg, 2007). Indeed, navigating the counselling and psychotherapy process can at times resemble a balancing act between the will to respect professional ethics while attempting therapeutic interventions judged to be in the best interest of clients. That is, the line between what is considered therapeutic versus what is considered inappropriate in session is not always clear, and we must at times cope with incomplete empirical evidence and ambiguous ethical guidelines. Supervision becomes critical in these contexts as people seek to consult with those who are experienced and who possess enough knowledge and expertise to guide them.

The use of touch as a therapeutic intervention is a primary example of such a balancing act. The literature on touch is peppered with varying opinions and contrasting empirical evidence that highlights the potential for both positive and detrimental effects on clients (Bonitz, 2008; Calmes, Piazza and Laux, 2013; Eyckmans, 2009; Horton, Clance, Sterk-Elifson and Emshoff, 1995; Hunter and Struve, 1998; Phelan, 2009; Stockwell and Dye, 1980). Furthermore, apart from prohibiting sexual contact with clients, ethical codes from the major regulating bodies generally fail to address the issue of touch altogether (Calmes, Piazza and Laux, 2013). The ambiguous literature and vague guidelines for conduct leave supervisors ill-equipped to address issues of touch interventions, despite the serious implications that can ensue for client wellness and for supervisee development.

According to Burkholder, Toth, Feisthamel and Britton (2010), the ambiguous state of the literature regarding touch is a reflection of the lack of clarity that exists in ethical codes and among counsellors and psychotherapy practitioners and supervisors themselves. In their research, they report that a large number of counsellor educators and supervisors are unclear about issues of touch and are anxious about addressing this topic with students. This uncertainty leads to supervisors inadequately addressing issues of touch, leaving supervisees, in turn, uncertain and anxious regarding this issue (Burkholder et al., 2010). Tune (2001) also reported therapist ambiguity regarding touch as relating directly to the lack of coverage of this topic during training, apart from advising against it.

It seems, therefore, that avoidance of touch issues may stem from a lack of understanding and concrete knowledge regarding the safe and effective use of touch. Enhanced by the risks associated with the possible misinterpretation of touch (Ramsdell and Ramsdell, 1994), supervisors may be opting to advise against it or may be avoiding the topic altogether. In taking this approach, however, supervisors risk fostering rigidity in their supervisees, who may attempt to avoid ethical violations with inflexible boundaries, putting their therapeutic relationships at risk and eliminating the possibility of the potential benefits of touch for their clients (Durana, 1998; Gutheil and Gabbard, 1998; Horton et al., 1995; Kertay and Reviere, 1993; Stenzel and Rupert, 2004; Westland, 2011). Furthermore, avoidance of touch issues in supervision fails to address misunderstandings regarding touch (Bonitz, 2008), potentially leading to physical and ethical transgressions. In either case, clients are left disadvantaged and at risk for potential harm.

Avoid as we may, the reality is that counsellors and psychotherapists do touch their clients (Holroyd and Brodsky, 1977; Pope,

Tabachnick and Keith-Spiegel, 1987; Stenzel and Rupert, 2004), and disconcerting findings suggest that when they do, they may neglect important steps in protecting client well-being (Stenzel and Rupert, 2004). The ultimate outcome of touch interventions is said to depend on a complex interplay of individual counsellor and client factors and on how the touch is introduced and addressed in session (Durana, 1998; Phelan, 2009; Westland, 2011; Zur, 2007). Effective supervision of touch, therefore, requires both knowledge of the safe and effective use of touch in therapeutic settings as well as knowledge of important factors relevant to the supervisory process and outcomes.

Knowing that touch between therapists and clients does happen and that it has the potential for positive or negative effects, the important question for supervisors is not whether supervisees should touch their clients (Kertay and Reviere, 1993), but rather: *How can we, as supervisors best supervise this act while developing supervisee technical competence, encouraging their self-reflection and ethical awareness, all the while ensuring client safety and protecting our own interests?* This may sound like a daunting task; however, my belief is that this can be accomplished relatively simply by reflecting on the use of touch from a process integration approach, geared towards fostering reflective practice, and protecting and monitoring the therapeutic alliance.

The next part of this chapter will present an approach to supervising touch based on a trans-theoretical alliance model to help guide supervisors in their ethical and respectful supervision of touch interventions. Although this model will be discussed in the context of supervising touch interventions specifically in counselling and psychotherapeutic contexts, it is my belief that it can be applied to the supervision of a variety of grey issues in a variety of mental health professions.

A Model for Supervising Touch

There are a myriad of theoretical articles providing recommendations on the ethical use of touch that serve to inform best practices for the use of touch in therapeutic settings. From this literature, we can clearly identify three important process factors that commonly are cited as determining factors in the safe and effective use of touch: (1) obtaining consent to touch; (2) explicating intent and expected effects of the touch intervention; and (3) aligning the touch intervention with the level of intimacy (Bonitz, 2008; Durana, 1998; Eykmans,

2009; Horton et al., 1995; Phelan, 2009; Suiter and Goodyear, 1985; Westland, 2011).

These three components can be mapped to Bordin's (1983) conceptualization of the working alliance, as a model to guide reflection and critical thinking processes regarding the use of touch interventions. Bordin conceptualizes the working alliance as having three main components: (1) mutual agreement on goals to be attained; (2) mutual agreement on the tasks to be undertaken to achieve these goals, both of which are grounded in; (3) a strong emotional bond. The emotional bond within the working alliance is said to be developed through shared emotional experiences which center "around feelings of liking, caring, and trusting that the participants share" (p. 36). Initially conceptualized in terms of therapeutic context (1979), Bordin later expanded his application of the model pan-theoretically to all contexts of the helping professions, including the supervisory context (Bordin, 1983).

The rationale for using a process supervision approach focused around the therapeutic alliance is grounded in the empirical research outlining the working alliance as a significant predictor of therapeutic outcome (Norcross and Wampold, 2011), and as one of the most important factors relating to positive outcomes of touch interventions (Horton et al., 1995).

Figure 1 provides a visual representation of the applications of Bordin's working alliance model to best practices of touch interventions. As depicted below, supervisors can guide a discussion of touch issues around the therapeutic alliance in order to safely and effectively address and monitor issues of touch.

When discussing issues of touch with supervisees, supervisors can explore each dimension for a clear understanding of the nature and effects of the touch used. In terms of mutually agreed-upon goals, supervisors can question supervisees about how the touch intervention was brought forth and negotiated into the alliance. Supervisors can invite supervisees to articulate the conversation that surrounded its alignment with therapeutic goals and how it was negotiated between the dyad.

With regards to agreement on tasks, supervisors can invite the supervisee to reflect on and articulate explicitly how the intervention was defined by both the counsellor and the client, and what boundaries were mutually set. The supervisee can also be invited to discuss the links that were made between the act of touching and the agreed upon goal(s).

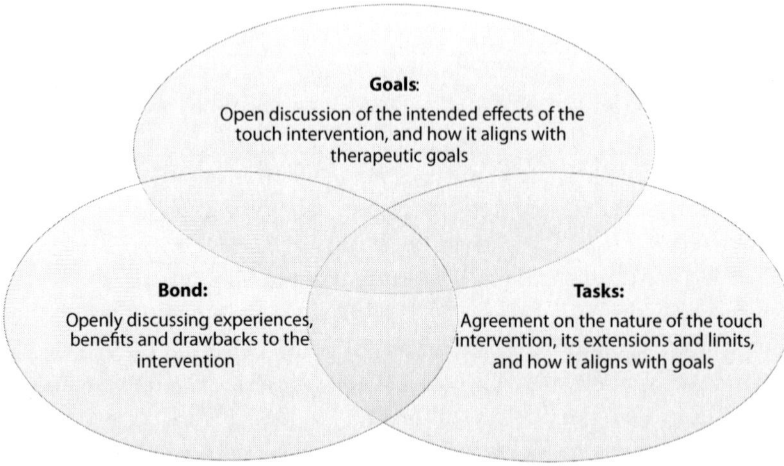

Figure 1. Applications of Bordin's alliance model to best practices of touch interventions

Finally, supervisors can invite their supervisees to articulate the discussion that was had between themselves and the client around the impact of the touch on their mutual feelings of trust, liking, and caring. This will allow supervisors to monitor bond fluctuations and the effectiveness of the intervention, while encouraging the supervisee to reflect on both their own feelings and that of their clients.

Inevitably, not all supervisees will have had these conversations openly with their clients, and some supervisees may even have blind spots or attempt to hide things from their supervisors (Daniels, 2000; Tune, 2001). Supervisors cannot be expected to have psychic powers, but Feltham (2000) asserts that much of what goes on in therapy can be detected by asking relevant questions and closely monitoring case details. If we are to act responsibly as supervisors, we must be aware of the possibilities of touch and inquire directly about it (Daniels, 2000). It is possible that some supervisees may be unaware that touching their clients may be impacting them or unaware of their own reasons for doing so, while others may be ridden with the anxiety of not knowing how to handle incidents of touch and, coupled with the fear of negative evaluation, may be avoiding bringing these issues to supervision (Feltham, 2000).

Encouraging reflections and considerations around the therapeutic alliance aligns with both supervisory and ethical obligations; they serve to foster competence, self-reflection and ethical awareness all the while ensuring client safety. Furthermore, this process approach simultaneously allows supervisors to address best practices in regards to touch in supervision. Inviting supervisees to openly discuss touch interventions around alliance factors will assure that consent is obtained, that the intervention was explicated, and that alignment of intervention with emotional intimacy was considered and open for discussion.

Of critical importance in this approach, however, will be the supervisor's own skills and qualities (Daniels, 2000). That is, the supervisor's own ability to develop a strong working alliance with their supervisee and to address touch issues with flexibility will bear great impact on their effectiveness in supervising touch. Furthermore, a supervisor's awareness of important notions to consider in using touch is of utmost importance. Supervisors may need to consider other important client and counsellor factors into the conversation around the therapeutic alliance. Client factors such as age, gender, culture, past experiences, presenting problem, and personality (Zur, 2007; Bonitz, 2008) have been found to differentially impact the effectiveness of touch. Moreover, a supervisee's own attitudes regarding touch and their awareness of the inherent power differentials regarding touch can also play a key role (Durana, 1998).

In summary, a process approach to supervision aimed at fostering self-awareness and ethical reflection while monitoring fluctuations in the therapeutic alliance can be an effective means to supervising touch issues between supervisees and their clients, regardless of approach or context. Inviting the supervisee to reflect on their use of touch and how this was negotiated around the alliance can serve as a non-threatening approach to fostering reflective practice within supervisees, all the while providing the supervisor with a window into the supervisee/client alliance dynamics, thus monitoring safe and effective use of touch.

Reflective practice involves taking a metacognitive perspective of one's interventions that lead to changes in one's future practice (Scaif, 2010). Encouraging supervisees and assisting them in reflecting on their actions and interventions regarding touch will likely lead to them developing this ability on their own (Scaif, 2010). In this regard, using a concrete model to guide this reflection will ultimately serve to diminish supervisee anxiety regarding issues of touch and possibly contribute to breaking the cycle of avoidance.

Touch Between Supervisor and Supervisee

Supervision is an emerging field of practice and much of our knowledge and understanding of supervision stems from extrapolations from the counselling and psychotherapy literature (Milne, 2006). Given the fact that issues of touch are still not fully understood, it is not surprising that a search for literature concerning touch in supervisory contexts has yielded no results. However, there is growing interest in identifying competencies and best practices in supervision (Borders, 2014; Watkins, 2014), and the calls from literature have echoed this need for supervisory guidelines regarding touch (Eyckmans, 2009; Hunter and Sturve, 1998). It is timely, therefore, that we begin to explore and address the safe and effective use of touch within the supervisory context. The following section is intended to begin this reflection; and the reflections will be guided by a critical look at the standards of professional supervision practice and informed by current existing literature on the use of touch in therapeutic contexts.

A review of ethics codes that relate to counselling and psychotherapy supervision parallel those of practitioners and provide little or no guidance concerning touch in supervision, besides the prohibition of sexual contact between supervisees and supervisors (ACA; AAMFT; CASC; CCPA; CPA). Documents outlining best practices in supervision serve to supplement codes of ethics and to outline evidence-based guidelines for applying identified competencies (Borders, 2014). The most cited and complete of these documents to date is the *Best Practices in Clinical Supervision,* developed by The Association for Counselor Education and Supervision (ACES, 2011), a division of the ACA. This document was developed to be applicable to a broad range of supervisory settings. The reflection will focus on two sections of this document, which in my opinion are of particular relevance to addressing touch in the supervisory context.

The first section of particular relevance is section five: "The Supervisory Relationship." Dedicated to fostering optimal supervisory relationships and alliances, this section highlights the supervisory working alliance as a key ingredient to the effectiveness of supervision (ACES, 2011). The literature on touch in therapeutic contexts points to a reciprocal relationship between touch and working alliance, highlighting the role of a strong working alliance in predicting efficacy of touch interventions (Horton et al., 1995; Zur, 2007), as well as the role of touch interventions in strengthening the alliance

(Burkholder et al., 2010; Calmes, Piazza and Laux, 2013; Durana, 1998; Kertay and Reviere, 1993; Smith, 1998). In light of the literature suggesting that the working alliance plays an equally important role for supervisory outcomes (Ladany, Ellis and Friedlander, 1999), it seems reasonable to assume that touch within a supervisory context can potentially have similar positive impacts if used responsibly and ethically. However, significant distinctions between alliances in supervision and in therapeutic contexts must be considered (Angus and Kagan, 2007). Specifically, although the supervisory working alliance is construed the same way as the therapeutic alliance, we must be sensitive to the fact that goals, tasks, and the nature of the bond within the supervisory working alliance differ significantly from that in therapeutic contexts (Angus and Kagan, 2007).

In terms of goals, touch in therapeutic contexts has been linked to healing outcomes for clients (Durana, 1998; Kertay and Reviere, 1993). In the supervisory context, however, supervisors are ethically prohibited from engaging in therapy with their supervisees (ACA; AAMFT; CASC; CCPA; CPA). The goals of supervision typically focus on learning and developmental outcomes rather than on becoming psychologically well (Bernard and Goodyear, 2014), as it is a prerequisite to the act of counselling and psychotherapy that supervisees be psychologically well when providing services (ACA; AAMFT; CASC; CCPA; CPA). From this standpoint, examples of appropriate alliance goals in the supervisory setting could include the supervisee learning the safe and effective use of touch techniques. Tasks related to accomplishing such goals could include the supervisor modeling touch behaviours and/or the supervisee attempting to experiment with such techniques. Supervisors and supervisees could process the use of the touch interventions, their possible benefits, and potential harmful effects, as well as each participant's experience of touch regarding how the technique was delivered. This is very different than touching clients with intent to heal. With regards to bond, touch in the therapeutic context is seen as a tool for strengthening the bond (Calmes, Piazza and Laux, 2013; Durana, 1998; Eykmans, 2009; Geib, 1998). For example, while a bond can naturally occur through accomplishing agreed-upon tasks, it has been suggested that touch can also strengthen bonds by providing clients with a reparative positive experience, by grounding them when working through difficult emotions, and by communicating acceptance, which establishes an atmosphere of safety and comfort when dealing with difficult material (Calmes, Piazza and Laux, 2013;

Durana, 1998; Eykmans, 2009; Geib, 1998). While some supervisors may choose to use touch as communication of acceptance, or even as a marker of support when supervisees become distraught, providing supervisees with reparative relational experiences may fall out of the scope of the supervisory mandate, leading supervisors down the slippery slope of dual relationships.

The second section of the ACES *Best Practices in Clinical Supervision* that is of particular relevance to the discussion of the use of touch in supervision is section eleven: "The Supervisor." This section is dedicated to the requirement of supervisors being competent themselves, and highlights themes of self-monitoring and self-reflection. Indeed, supervisors need to lead by example. Encouraging supervisees to be self-reflective in their own practice necessarily requires that supervisors be able to do the same (Bernard and Goodyear, 2014). Any use of touch in supervision should be considered from all possible angles and take into consideration the particularities of the supervisory context. Supervisors must be able to reflect themselves on their own intentions, their values, as well as their intended use of touch with supervisees. As is recommended in therapeutic contexts where touch is used, supervisors must be open to discussing with supervisees the uses of touch and how it aligns with the mutually agreed-upon goals of the supervision or with the strengthening the bond between the two (Westland, 2011; Kertay and Reviere, 1993).

Supervisors who choose to use touch in supervision must also be aware of the inherent ingredients of the supervisory process that may influence the effectiveness of the touch interventions. The first of these is acknowledging the distinct power differentials that exist in the supervisory relationship which relate to the evaluative and gatekeeping role of supervisors (Bernard and Goodyear, 2014). That is, in supervisory contexts, even the most collaborative and humanistic approaches cannot override the fact that supervisors must ultimately evaluate the competence of their supervisees. This very fact has been found to intimidate supervisees and limit their ability to challenge supervisors or disclose concerns (Lawton, 2000). Supervisors must be extra sensitive to this factor in their management of touch between themselves and their supervisees, for supervisees may feel unable to express discomfort. Furthermore, the context may be inappropriate to process this discomfort even if it were to be expressed.

Second, supervisors need to be aware of possible role confusion and sensitive to its potential negative impacts. Fluctuating needs of supervisees require that supervisors alternate between the different roles of consultant, teacher, evaluator, and counsellor within session, which may lead to role confusion for both parties (Bernard and Goodyear, 2014). Outside of session, role confusion is also at greater risk as supervisors may find themselves in unavoidable dual relationships with their supervisees (who may also be students in the department in which they teach or mutual members of a professional organization) (Navin and Beamish, 1995). The lack of clarity concerning roles and relationships may render supervisory relationships more susceptible to boundary violations (Sherry, 1991). That is, supervisors and supervisees may inadequately address touch in these relationships, as they may be viewed by both parties as more collegial in nature.

Lastly, supervisee factors of age, gender, culture, past experiences, and personality (Zur, 2007; Bonitz, 2008) may have an important impact on the effectiveness of touch in supervision, and supervisors are advised to explore these with their supervisees before considering the use of touch.

Recommendations

In light of the previously highlighted differences between the therapeutic and supervisory processes, the following is a summary of recommendations for supervisors to consider for the safe and effective use of touch between themselves and their supervisees:

1. Supervisors should be keenly aware of the differences existing between therapeutic alliances and supervisory alliances. Before engaging in touch with their supervisees, supervisors need to understand both the potential benefits and damages of touch, as well as have a good understanding of the process of supervision and its functional role. Mutually agreed-upon touch interventions must be aligned to supervisory goals and should be openly discussed and debriefed for its effects on both dyad members.

2. Supervisors should be cognizant of the power differential inherent in the supervisory relationship, and factor in the possibility that this may limit the free will of their supervisees and their ability to express concern regarding the use of touch in supervision.

3. Supervisors must lead by example. In order to foster self-awareness and reflective practice, supervisors must be able to demonstrate this skill as it applies to their own practice. This will not only

serve as a modeling function for supervisees but will assist supervisors in developing reflective supervisees.

4. Supervisors should be well versed on the differential impacts of individual factors on the effectiveness of touch within the supervisory context as they relate to age, gender, culture, past experiences, and personality.

Conclusion

It is evident from the multitude of theoretical standpoints and conflicting findings in the literature that touch is a complex issue to address in therapy, which renders it even more complex to supervise effectively. What is clear is that avoidance of these issues does not make them go away, but rather perpetuates a culture of anxiety and misunderstanding surrounding touch, putting supervisors and supervisees at risk for misusing touch and potentially harming their clients. Grounded in the literature on the best practices of touch in therapeutic settings and on the nature of the supervisory process and outcomes, this chapter provides a concrete model for accompanying supervisors in their supervision of touch issues between supervisees and their clients in an ethical, competent, and sensitive way. Furthermore, this chapter begins the conversation on touch between supervisors and supervisees which, to date, has remained absent from the literature, providing some preliminary guidelines on best practices for supervisor/supervisee touch.

References

American Counselling Association. (2014). Code of ethics. Retrieved from http://www.counseling.org/knowledge-center/ethics

American Association for Marriage and Family therapy. (2012). Code of ethics. Retrieved from http://www.aamft.org/iMIS15/AAMFT/Content/legal_ethics/code_of_ethics.aspx

Angus, L. and Kagan, F. (2007). Empathic relational bonds and personal agency in psychotherapy: Implications for psychotherapy supervision, practice, and research. *Psychotherapy: Theory, Research, Practice, Training, 44*, 371–377. doi:10.1037/0033-3204.44.4.371

Association for Counselor Education and Supervision, Task force on Best Practices in Clinical Supervision. (2011). *ACES Task Force Report*. Retrieved from https://www.wcu.edu/WebFiles/PDFs/AcesBestPractices.pdf

Barnett, J. E., Cornish, J. A. E. and Goodyear, R. K. and Litchenberg, J. W. (2007). Commentaries on the ethical and effective practice of clinical supervision. *Professional Psychology: Research & Practice, 38,* 268–275. doi: 10.1037/0735-7028.38.3.268

Bernard, J. M. and Goodyear, R. K. (2014). *Fundamentals of clinical supervision* (5th ed.). Upper Saddle River: Pearson.

Bonitz, V. (2008). Use of physical touch in the "talking cure": A journey to the outskirts of psychotherapy. *Psychotherapy Theory, Research, Practice, Training, 45,* 391–404. doi: 10.1037/a0013311

Borders, D. (2014). Best practices in clinical supervision: Another step in delineating effective supervision practice. *American Journal of Psychotherapy, 68,* 151–162. Retrieved from https://www.researchgate.net/publication/264352000_Best_Practices_in_Clinical_Supervision_Another_Step_in_Delineating_Effective_Supervision_Practice

Bordin, E. S. (1979). The generalizability of the psychodynamic concept of the working alliance. *Psychotherapy: Theory, Research, and Practice, 16,* 252–260. doi: 10.1037/h0085885

Bordin, E. S. (1983). A working alliance model of supervision. *The Counseling Psychologist, 11,* 35-42. doi: 10.1177/0011000083111007

Burkholder, D., Toth, M., Feisthamel, K. and Britton, P. (2010). Faculty and student curricular experiences of non-erotic touch in counseling. *Journal of Mental Health Counseling, 32,* 168–185. Retrieved from http://web.a.ebscohost.com/ehost/detail/detail?sid=2bf81a23-eb21-4f56-ba42-d0ca3787803b%40sessionmgr4004&vid=0&hid=4209&bdata=JnNpdGU9ZWhvc3QtbGl2ZQ%3d%3d#db=a9h&AN=50364716

Calmes, S. A., Piazza, N. J. and Laux, J. M. (2013). The use of touch in counseling: An ethical decision-making model. *Counseling and Values, 58,* 59-68. doi: 10.1002/j.2161-007X.2013.00025.x

Canadian Association for Spiritual Care. (2006). Common code of ethics for chaplains, pastoral Counselors, pastoral educators and students. Retrieved from http://www.spiritualcare.ca/page.asp?ID=46

Canadian Counselling and Psychotherapy Association. (2007). Code of ethics. Retrieved from http://www.ccpa-accp.ca/en/resources/codeofethics/

Canadian Psychological Association. (2000). Canadian code of ethics for psychologists. Retrieved from http://www.cpa.ca/aboutcpa/committees/ethics/codeofethics

Daniels, J. (2000). Whispers in the corridor and kangaroo courts: The supervisory role in mistakes and complaints. In B. Lawton and C. Feltham (Eds.), *Taking supervision forward: Enquiries and trends in counselling and psychotherapy.* London: Sage.

Durana, C. (1998). The use of touch in psychotherapy: Ethical and clinical guidelines. *Psychotherapy, 35,* 269–280. doi: 10.1037/h0087817

Eyckmans, S. (2009). Handle with care: Touch and a therapeutic tool. *Gestalt Journal of Australia and New Zealand, 6,* 40–53. Retrieved from http://search.informit.com.au/ documentSummary;dn=349896602090634;res=IELHEA

Falendar, C. A. and Shafranske, E. P. (2007). Competence in competency-based supervision practice: Construct and application. *Professional Psychology: Research & Practice, 38,* 232–240. doi: 10.1037/0735-7028.38.3.232

Feltham, C. (2000). Counseling supervision: Baselines, problems and possibilities. In B. Lawton and C. Feltham (Eds.), *Taking supervision forward: Enquiries and trends in counselling and psychotherapy.* London: Sage.

Gutheil, T. G. and Gabbard, G. O. (1998). Misuses and misunderstandings of boundary theory in clinical and regulatory settings. *American Journal of Psychiatry, 155,* 409–414. Retrieved from http://search.proquest.com/docview/220469990/abstract?accountid=14701

Holroyd, J. C., and Brodsky, A. (1977). Psychologists' attitudes and practices regarding erotic and nonerotic physical contact with patients. *American Psychologist, 32,* 843–849. doi: 10. 1037/0003-066X.32.10.843

Horton, J. A., Clance, P. R., Sterk-Elifson, C. and Emshoff, J. (1995). Touch in psychotherapy: A survey of patients' experiences. *Psychotherapy, 32,* 443–457. doi: 10.1037/0033-3204.32.3.443

Hunter, M. and Struve J. (1998). Challenging the taboo: Support for the ethical use of touch in psychotherapy with sexually compulsive/addicted clients. *Sexual Addiction & Compulsivity, 5,* 141–148. doi: 10.1080/10720169808400156

Kertay, L. and Reviere, S. L. (1993). The use of touch in psychotherapy: Theoretical and ethical considerations. *Psychotherapy, 30,* 32–40. doi: 10.1037/0033-3204.30.1.32

Lawton, B. (2000). 'A very exposing affair': explorations in counsellors' supervisory relationships. In B. Lawton and C. Feltham (Eds.), *Taking supervision forward: Enquiries and trends in counselling and psychotherapy.* London: Sage.

Milne, D. (2006). Developing clinical supervision research through reasoned analogies with therapy. *Clinical Psychology and Psychotherapy, 13,* 215–222. doi: 10.1002/cpp.489

Navin, S. and Beamish, P. (1995). Ethical practices of field-based mental health counselor supervisors. *Journal of Mental Health Counseling, 17,* 234–255. Retrieved from http://web.b.ebscohost.com.proxy.bib.uottawa.ca/ehost/detail/detail?sid=eb4636b9-ef50-40fa-8359 a5f776da6ce3%40sessionmgr112&vid=0&hid=110&bdata=JnNpdGU9ZWhvc3QtbGl2ZQ%3d%3d#db=a9h&AN=9509242613

Norcross, J. C. and Wampold, B. E. (2011). Evidence-based therapy relationships: Research conclusions and clinical practices. *Psychotherapy, 48,* 98–102. doi:10.1037/a0022161

Phelan, J. E. (2009). Exploring the use of touch in the psychotherapeutic setting: A phenomenological review. *Psychotherapy Theory, Research, Practice, Training, 46,* 97–111. doi: 10.1037/a0014751

Pope, K. S., Tabachnick, B. G. and Keith-Spiegel, P. (1987). Ethics of practice: The beliefs and behaviors of psychologists as therapists. *American Psychologist, 42,* 993–1006. doi:10.1037/0003-066X.42.11.993

Ramsdell, P. S. and Ramsdell, E. R. (1994). Counselor and client perceptions of the effect of social and physical contact on the therapeutic process. *Clinical Social Work Journal, 22,* 91–104. doi: 10.1007/BF02190288

Scaif, J. (2010). *Supervision the reflective practitioner: An essential guide to theory and practice.* New York: Routledge.

Sherry, P. (1991). Ethical issues in the conduct of supervision. *The Counseling Psychologist, 19,* 566–584. doi: 10.1177/0011000091194006

Smith, E. (1998). Traditions of touch in psychotherapy. In E. Smith, P. R. Clance and S. Imes (Eds.), *Touch in psychotherapy, theory, research and practice* (pp. 3–15). London: Guilford Press.

Stenzel, C. L. and Rupert, P. A. (2004). Psychologists' use of touch in individual psychotherapy. *Psychotherapy: Theory, Research, Practice, Training, 41,* 332–345. doi: 10.1037/0033-3204.41.3.332

Stockwell, S. R. and Dye, A. (1980). Effects of counselor touch on counseling outcome. *Journal of Counseling Psychology, 27,* 443–446. doi: 10.1037/0022-0167.27.5.443

Suiter, R. L., and Goodyear, R. K. (1985). Male and female counselor and client perceptions of four levels of counselor touch. *Journal of Counseling Psychology, 32,* 645–648. doi:10.1037/0022-0167.32.4.645

Tune, D. (2005). Is touch a valid therapeutic intervention? Early returns from a qualitative study of therapist's views. *Counselling and Psychotherapy Research, 1,* 167–171. doi:10.1080/ 1473 3140112331385020

Watkins, C. E., Jr. (2014). Clinical supervision in the 21st century: Revisiting pressing needs and impressing possibilities. *American Journal of Psychotherapy, 68,* 251–272. Retrieved from http://web.a.ebscohost.com.proxy.bib.uottawa.ca/ehost/pdfviewer/pdfviewer?sid=139ddc7c-f294-48e3-b692-6c3675dc0573%40sessionmgr4005&vid=2&hid=4207

Westland, G. (2011). Physical touch in psychotherapy: Why are we not touching more? *Body, Movement and Dance Psychotherapy, 6,* 17–29. doi: 10.1080/17432979.2010.508597.

Zur, O. (2007). Touch in therapy. In *Boundaries in psychotherapy: ethical and clinical Implications* (pp. 167–185). Washington, DC: American Psychological Association.

PART IV

THE ETHICS OF TOUCH AND CONCLUSION

Chapter XII

The Ethics of Therapeutic Touch

Marilyn Guindon, Reesa Packard and Natalie Charron

Many would be surprised to learn that Freud (Kertay and Reviere, 1993) touched his patients when seeing them in his office, sometimes massaging or touching their heads. Also, Freud utilized touch to help clients through states of mutism (Breuer and Freud, 1957). Because Freud wanted psychoanalysis to be perceived as an objective science, he stopped his practice of therapeutic touch (Phelan, 2009). More precisely, psychoanalysts started being wary that all use of therapeutic touch would contaminate the transferential relationship and, as such, result in interference in the analysis as well as in the risk that personal gratification of the therapist could outweigh the client's psychological well-being (Johnston and Farber, 1996; Langs, 1982). Bowlby's (1969) research, however, found that all human beings were naturally pulled towards intimate contact with other human beings. Ainsworth (1989), thanks to her experimental research titled "Strange Situation," was able to demonstrate that physical contact, particularly between mother and child, was essential for the child's attachment and emotional connection. In addition, Ainsworth illustrated that the child's attachment experience to their primary caregivers had an impact on their adult life. During this same period, psychotherapy, which to that point was mainly centered on the theoretical and technical aspects of the profession, was also turning its attention to professional ethics, as well as risk management (Bonitz, 2008). Thus, the influence of Freudian doctrine (Durana, 1998), the presence of cultural taboos in a particular era (Cohen, 1987; Heller, 1997), the scientific knowledge on the importance of human contact for optimal development (Ainsworth, 1989), and the legal and ethical considerations (Bonitz, 2008) are at the heart of this professional discussion regarding the use of touch in the practice of psychotherapy.

A Question of Personal Boundaries

Therapeutic or Not?

As proposed by Montagu (1986), touch is the leader of all senses, as it molds human relations: It is the foundation of human experience. The kinetic sensations and the perceptions of touch are generally associated with what is nourishing, calming, and even healing (Warnecke, 2011). However, touch can also be erotic, a symbol of desire and passion, which can lead to exploitation (Warnecke, 2011). Juhan (1987) captures the power of therapeutic touch well when he suggests that if we touch the surface of the body, we stir the depths. These depths have the potential to evoke powerful transferences, to provoke limbic resonances between the therapist and the client that open to resonance, synchronicity, and empathy with the client experience (Warnecke, 2011). This experience, while shared, implies that the therapist not be a neutral screen, but rather a mirror within which the client's projective symbolism can operate (Warnecke, 2011). Obviously, these boundaries are also determined by defined parameters such as the limits of the relationship which, within the framework of psychotherapy, is influenced by professional conduct and ideology, the therapy contract, the client consent to therapy, as well as the therapeutic context (Norris, Gutheil, and Strasburger, 2003). As a result of this, how can we distinguish between a therapeutic touch and one that has an ulterior motive? Warnecke (2011) offers an extensive answer to this question by specifying:

> Both therapist and client seek to make contact with the other. Both contribute their unique subjectivities to the therapeutic alliance which is necessarily co-created. Such a shift from one-person to two-person psychology involves moving the therapist's position from interpreting observations or administrating treatment to participation, facilitation, and co-creation. This allows therapist and client to explore but also to cross psychological boundaries

> creatively within the protected formal frame of the therapeutic relationship. Projections, transferences and projective identifications form a seamless continuum which exists side by side with the real relationship and the personalities of the therapist and client. Boundaries between benign and malignant touch are determined by context, intention, and meaning. (pp. 235–236)

Now, it is incumbent on us to distinguish transgression from violation of boundaries to further nuance these statements. As such, when the psychotherapist transgresses the client's boundaries, he deviates from his professional conduct. Yet the consequences of his conduct are generally without danger for the clients and may even be considered an asset in treatment (Norris and al., 2003). For example, the psychotherapist can help a client with reduced mobility to move by holding his arm or accompanying him in a taxi to the hospital when he is in crisis. However, when there is a violation of a client's boundaries, irreparable damage occurs and/or exploitation of the client for the psychotherapist's personal gain (Norris and al., 2003). To this effect, some of the most commonly cited unethical behaviours are: having sexual relations with a client, and having a financial relation with a client or authority in their regard.

While other authors (Gabbard, 1994, 1996; Simon, 1989; Strasburger, Jorgenson and Sutherland, 1992) agree that certain transgressions of boundaries can contribute in a positive way to treatment, the repetitiveness of these transgressions and the ambiguity it communicates to the client has the potential to generate more serious problems, thus leading to a violation of boundaries that can put an end to the therapeutic relationship. Gabbard (1994) indicates that the transgression of a single boundary without catastrophic consequences opens the client to future instances of its transgression. As such, many studies have demonstrated that the majority of boundary violations were preceded by minor transgressions (Bory and Pope, 1989; Koocher and Keith-Spiegel, 1998; Lakin, 1991; Rutter, 1989; Sonne, 1994; St. Germaine, 1996; Strasburger et al., 1992). We must be prudent, says Zur (2007), as "a sequential relationship or correlation does not equal causality" (p. 52). Therefore, a transgression of boundaries does not necessarily generate a violation of boundaries, although often transgression does precede violation.

Current Context

With the development of neuroscience, research on trauma, as well as child development, abstinence as a rule surrounding therapeutic touch deserves re-evaluation (Galton, 2006; Orbach and Carroll, 2003). Relatively recent surveys among psychotherapists reveal that nearly 87% of them touch their clients (Tirnauer, Smith, and Foster, 1996), 85% hug their clients (Pope, Tabachnick and Keith-Spiegel, 1987), and 65% consider this type of touch, while not a technique in the strictest sense, an essential asset to talk psychotherapy (Schultz, 1975). A more recent study among social workers, from Strozier, Krizek and Sale (2003), illustrated that 95% of them had touched their clients at least once, and that nearly a third (29%) did it frequently or very frequently. In addition, the most common types of contacts were shaking hands, a touch of the forearm, the shoulder or neck, as well as the embrace, a hug with a tap of the back for encouragement (Strozier et al., 2003). In another survey of 200 psychologists, it was revealed that shaking hands, touching the shoulder, the arm and the thigh, as well as the hug, were rarely perceived as an issue of professional conduct and were not systemically associated with suggestive or sexual behaviours (Stake and Oliver, 1991). Others still (Borys and Pope, 1989; Johnston and Farber, 1996; Pope, Tabachnick, and Keith-Spiegel, 1987) report that most psychotherapists shook hands and exercised therapeutic touch in an appropriate manner. Finally, Wilson (1982) was able to elucidate that most psychotherapists weren't entirely closed off regarding the use of touch, but rather that they were reticent to admit its use within their practice. On this, the following section explores the fears psychotherapists have concerning the use of therapeutic touch as it relates to professional ethics.

What Professional Ethics Say

First, it is important to understand that most associations and professional colleges in the field of mental health and, more specifically, psychotherapy, do not forbid or condemn the use of touch within the therapeutic relationship or treatment (AAMFT, 2012; CCPA, 2007; CPO, 2005; OPQ, 2014; Phelan, 2009). However, the fear of being accused of sexual misconduct remains the principal reason many psychotherapists prefer to abstain from exercising any form of touch (Phelan, 2009; Norris and al., 2003). In a survey of psychologists, 44%

reported that at least one of their clients had mentioned having physical contact with their prior psychotherapist (Stake and Oliver, 1991). In 2006, client complaints concerning the occurrence of sexual contact with their psychotherapist topped the list (Phelan, 2009). Pope et al. (1987) found that between 1975 and 1985, 9–12% of male psychotherapists and 2–3% of female psychotherapists had sexual contact with their clients. During the 1990s, these percentages dropped dramatically, essentially due to the ethical and legalist shift within the profession. While these were self-reported statistics, by 1994, between 0.9% to 3.6% of male psychotherapists and 0.2% to 0.5% of female psychotherapists report having sexual contact with their clients (Pope, 2001). The culture of risk management, while serving to reduce sexual contact between psychotherapists and their clients, also led them to avoid touch, despite the benefits of certain interventions that involve touch (Zur and Nordmarken, 2009). In sum, mental health professionals stopped all discussion surrounding the theme of therapeutic touch for fear of being accused of sexual misconduct by their clients (Stenzel and Rupert, 2004: Harrison et al., 2012). This risk management left expectations that psychotherapists do not touch their clients and do not discuss touch with them. Deemed inappropriate, the fact that there is a failure to exercise therapeutic touch reaffirms the belief that a good practice does not include it. Yet the benefits of touch are the subject of many studies that support its therapeutic value (Bonitz, 2008; Harrison et al., 2012; Phelan, 2009; Westland, 2011; Zur, 2007).

The Benefits of Therapeutic Touch

According to Imes (1998), if we only talk and refuse to touch our client, it is possible we are missing an opportunity to allow them to express certain emotions by this avenue that, sometimes, is the only one within their reach. From a client's point of view, being touched by their psychotherapist is generally well received if the contact is appropriate, and not abusive, coercive, or sexual (Phelan, 2009). Touch can have a reassuring effect on the client, resulting in being more engaged in therapy (Boderman et al., 1972), perceiving the psychotherapist as being someone trustworthy (Hubble and al., 1981), and opening up with greater ease (Pattison, 1973). Hunter and Struve (1998) were able to demonstrate that touch managed to establish, maintain, and even deepen the therapeutic alliance between the client and the psychotherapist. Along these lines, these authors also proposed that therapeutic touch allowed certain clients to remain in the moment, promoted accessibility, exploration and the resolving of certain emotional experiences. Additionally, therapeutic touch gave clients a sense of trust and security and offered to the psychotherapist the possibility to promote touch with clients as being a healthy aspect of interpersonal relationships that can be lived in a positive way outside of the therapeutic context (Hunter and Struve, 1998). According to Bonitz (2008), therapeutic touch allows the psychotherapist to help the client work through certain resistances, to communicate a symbolic acceptance as we are, and to offer an anchoring opportunity. Other authors push this further and present touch as being humanizing and nourishing for the clients (Phelan, 2009). The benefits of including touch within the therapeutic relationship can be enriching (Westland, 2011) for many reasons, including (1) containing, providing security, reorienting, and bringing back to reality traumatized clients; (2) calming and stabilizing the autonomous nervous system and fostering a sense of security via the non-verbal, as well as restoring the psycho-physiological system of clients deregulated on this level; (3) inciting an emotional reaction, an abreaction, a relief, to reduce the resistance of certain clients; (4) symbolic re-parenting; (5) exploring and differentiating the notions of aggression and pleasure; (6) increasing and encouraging the circulation of body energy and deepen breathing; (7) deepening the experience in the therapeutic relationship while maintaining the client's attention and encouraging the exploration of his past as well as discovering and exploring the relational needs of the client, and; (8) encouraging authenticity in the therapeutic relationship and feeling of acceptance, allowing for the emancipation and natural and spontaneous expression of the client. If the benefits of therapeutic touch are many, it is certainly not enough to justify its use within the psychotherapy framework. Many other factors are at the heart of the ethical considerations in exercising touch that is meant to be therapeutic, notably, the psychotherapist who exercises touch, the client on whom this intervention is effected, and the context in which touch is used.

How to Decide to Touch or not to Touch?

Who Touches?

Within the scientific documentation on therapeutic touch, variables linked to the psychotherapist, cultural considerations, personal, theoretical, and social variables are systemically put forth (Bonitz, 2008; Gutheil and Gabbard, 1993; Gutheil and Brodsky, 2008; Harrison et al., 2012; Norris et al., 2003; Phelan, 2009; Wen-Shing, 1999; Zur, 2007).

As for the cultural dimension, Phelan (2009) informs us that the psychotherapist's culture of origin has an impact on their use or non-use of touch within therapy. The psychotherapists of Asian origin are not inclined to touch and tend to want to avoid open demonstrations of affection as well as prolonged physical contact (Phelan, 2009). Psychotherapists of German origin come from a culture of non-touch. However, it is common among psychotherapists of Quebec, Italian, or Mediterranean origins to be more comfortable with physical contact. Smith and Fitzpatrick (1995) demonstrated that among Montreal psychotherapists, to kiss both cheeks was a common practice within the therapeutic context and that this behaviour was not a systemic communication of sexual messages or implied sexual tension within the therapeutic relationship.

The psychotherapist's gender, age, and past experiences also influence the use of touch within clinical practice (Zur, 2007). For example, being a woman (Harrison et al., 2012; Phelan, 2009; Milakovich, 1998; Stenzel and Rupert, 1994; Stozier et al., 2003), being older (Williams, 1997), and having had a previous positive experience of touch within a psychotherapeutic practice increase the probability that this intervention be used (Milakovich, 1998). Zur (2007) proposes that the sexual orientation of the psychotherapist can have an impact on their utilization of touch within therapy, without, however, presenting supporting data to that effect. Norris et al. (2003) outline distinct factors related to psychotherapists who exercise therapeutic touch in a non-helpful, possibly harmful, way within the context of psychotherapy. These include psychotherapists who (1) experience a crisis in their lives (e.g., marital conflict); (2) live life transitions (e.g., retirement, job change); (3) are ill; (4) experience loneliness; (5) have feelings of arrogance (e.g., feel they are above the law), of shame (e.g., be reticent to consult in case of doubt) or envy towards their clients (e.g., a counter-transference related to beauty, youth, intelligence, and the status of the client); (6) have difficulty putting limits in place within the therapeutic relationship; and (7) refuse to admit to problems within the therapeutic relationship (e.g., are in denial).

Again, from a psychotherapist's personal disposition perspective, the attitudes and beliefs that one holds regarding therapeutic touch are central in understanding the underlying motivation to utilize or not utilize touch. Harrison et al. (2012) specify that psychotherapists who use and value therapeutic touch tend to consider this intervention as being a way to meet a client's need, trust themselves to its appropriate use, are less preoccupied with the risks, and mostly use it to help provide clients access to their emotions. Bonitz (2008) indicates that psychotherapists who utilize touch tend to consider the potential to heal, to repair, that this intervention holds. Alternatively, psychotherapists who do not utilize or value touch within their practice tend to see this intervention in a negative way. Touch is perceived by them as being a source of gratification of the needs of the psychotherapist; touch is vulnerable to misinterpretation and may encourage a breach of boundaries on behalf of the client, to say nothing of the fact that this intervention may be simply intolerable for the psychotherapist (Harrison et al., 2012). Gutheil and Brodsky (2008) insist on the views psychotherapists have in regards to their own mental health. They indicate that the more the psychotherapist remains open and conscious of their vulnerabilities as well as their emotions towards their client(s), the more they will be equipped to manage the boundaries within the therapeutic relationship. In sum, the majority of psychotherapists who utilize touch specify that it is used with circumspection, mentioning the importance that the need to be touched originates from the client and that there is clear and informed consent (Harrison et al., 2012).

In regards to theoretical considerations related to the psychotherapist, the scientific documentation reveals that humanistic practitioners (30%), in contrast to those of a psychoanalytical approach (6%), tend to believe that therapeutic touch enables their clients to progress (Bonitz, 2008; Pope et al., 1987). Similarly, humanistic practitioners are more naturally inclined to utilize touch in light of the training and supervision they received that encouraged its practice (Bonitz, 2008). Other renowned practitioners-researchers, of somatic approach (body sensations) or attachment-centred approach (relational), also use touch within their practice in order to increase their client's general well-being. For example, Levine (2008) used somatic experiencing to resolve what he referred to as "combat responses, submission, and avoidance" (p. 27). This approach uses touch in a progressive manner; for example, by the pressure applied between hands or the backs of two partners, the client may learn the distinction between aggression and the feeling of empowerment. In assisting the client, the psychotherapist may also touch their client to improve or intensify the bodily sensations they feel (Levine, 2008). It is noted that professional ethics, in matters of touch with a therapeutic goal, is similar to that of verbal therapy, in that they invite the therapist to (1) evaluate if touch is an appropriate intervention for the client; (2) obtain informed consent on the client's behalf before using any form of touch; (3) monitor, in continuous manner, the client consent to all interventions exercised to ensure that touch remains a desired and appropriate intervention, according to and for the client; (4) acknowledge and respect the client's refusal to be touched; (5) be sure of their own capacity (competency and self-awareness) to use therapeutic touch; and (7) not engage in genital and/or sexual or sexually stimulating touch (Jaffy, 2002). Health care professionals have to respect the meaning the experience of touch reveals for the client, while remaining mindful of the imbalance of power that exists within the therapeutic relationship. In addition, we have the responsibility to develop our self-knowledge and self-awareness, as therapists, in order not to use touch in an inappropriate way that otherwise transgresses or violates the interpersonal boundaries of the therapeutic relationship.

Within couple counselling, Johnson (2008) indicates that touch allows for calming and comforting and can also prove essential, for example, when one of the clients finds themselves in a state of "primal panic" (p. 31); that is, that they feel great distress as a result of a rupture in connection with the relationship which results in the activation of their attachment system. In the Petrella and Rovers chapter, partners are encouraged and helped to find ways to reconnect through touch. Finally, cognitive-behaviourist psychotherapists can occasionally use touch to the extent that this intervention is context-appropriate, for example, within exposure therapy (progressive desensitization technique to an anxiety provoking stimulus) to which they have the informed consent of the client, which implies the treatment objectives were discussed and the expectations clarified (Gutheil, 2005). In such a therapeutic context, no precise guidelines exist as to the professional ethics of touch. It is only by extrapolating certain principles from the ethics code of the American Association for Marriage and Family Therapy (AAMFT, 2015), such as the principle of responsibility towards the client, as well as that of professional competence and integrity, that we may understand how certain rules may be applied to therapeutic touch within couple therapy. For example, a couples therapist has to obtain informed consent from clients, respect their rights, avoid all forms of conflict of interest, exploitation, and/or sexual contact with clients, and ensure a high degree of competence in exercising their profession. For the most part, it is the professionals from the field of body therapies (e.g., massage therapists, physiotherapists) who have developed an ethics code as well as precise guidelines as to the use of touch in a therapeutic relationship context. The practice of psychotherapy would benefit from the development of specific norms and guidelines in the use of therapeutic touch.

Gutheil and Brodsky (2008) argue that a change has occurred in the last 20 years in terms of psychotherapist expectations and professional behaviours that contributed to interactions within therapy to become more formalized. Now educated and formed within a society that promotes detachment in interpersonal relationships, this generation of psychotherapists is more reticent to transgress client boundaries (Gutheil and Brodsky, 2008). Yet we must equally consider the main variable of the client before taking any position related to the use of therapeutic touch.

Who Is Touched?

Horton et al. (1995) found that among a sample of 231 participants to psychotherapy, 69% of them perceived touch as being a means to deepen the feelings of connection, trust, and openness towards the psychotherapist. Furthermore, for 47% of clients, when touch used by their psychotherapist was perceived as sincere, had a tendency to communicate acceptance and positive regard of self, which had a ricochet effect of increasing their self-esteem (Horton et al., 1995). Many researchers (Knable, 1981; Levitan and Johnson, 1986; Lynch, 1978) specify that the response to touch from a client rests on an interaction of various factors, including the perceived intention, the expectations, and prior experiences of the client. Others propose that touch is generally beneficial for the client when (1) the client feels in control of the touch, and when the intervention meets the client's needs and not the needs of the psychotherapist (Geib, 1998; Horton et al., 1995); (2) touch is coherent with the difficulties the client presents (Horton et al., 1995); and; (3) it is possible to discuss the intervention (Geib, 1998; Horton et al., 1995). Bonitz (2008) adds that touch can also be perceived as favourable when the client has a positive impression of the therapeutic alliance, when she/he feels taken care of and close to the caregiver, and when the therapeutic relationship is sufficiently safe to explore psychological issues with him/her.

The survivors of various traumas were the subject of numerous studies on the efficacy and the pertinence of touch in the improvement of their symptomatology. Although Lawry (1997) states that touch should never be used with this clinical population, essentially due to the risk of activation or reactivation of traumatic memories, the studies that support the opposite are many. For example, Corness (1997) reports that once a therapeutic alliance forms, "The use of touch will broach, evoke, and possibly correct the experiences and distortions related to the abuse" (p. 33). He also specifies that when the need is manifested by the client and they consent freely and without coercion, therapeutic touch, even among trauma survivors, can be healing (Cornell, 1997). Finally, Imes (1998) indicates that a therapeutic touch respecting the boundaries of clients who survived various abuses and trauma offers them an opportunity to live an experience of touch that is non-abusive within a relationship in which they can learn to impose healthy limits, which positively contributes to their self-confidence and self-esteem.

Along these same lines, as proposed by Harrison et al. (2012), the use of therapeutic touch is also dependent on the level of distress experienced by the client, the extent of isolation, gender, behaviour in session, as well as their clinical presentation (e.g., client having boundary issues or not). Zur (2007) adds that touch is also dependent upon several factors that are essential elements to consider in the use of touch with a therapeutic goal. These include the culture of origin, the country of origin, the level of acculturation, the language of use, and language capacity, history (e.g., the presence of psychological, physical and/or sexual abuse), the presenting issue, the current and past psychological state, the socio-economic status, the living conditions, personality type, the presence of a personality disorder, sexual orientation, the spiritual beliefs and practices, the social support network, the state of health and mobility, the prior experience of therapy, and the general attitude towards therapy.

In addition to these variables, it becomes inevitable that the contraindications of the use of therapeutic touch, among certain clients, must also be the subject of discussion.

Contraindications

Research on the use of touch within the context of psychotherapy demonstrates that people having borderline traits or borderline personality disorder (BPD) (APA, 2013) will often have issues with interpersonal boundaries (Bonitz, 2008; Glickhauf-Huhues and Chance, 1998). Gurtheil (2005) indicates that those suffering from BPD tend to fuse and become demanding towards their psychotherapists. In addition, as these clients often present with a history of trauma, they tend to adopt a victimized position that is susceptible to mobilize feelings of helplessness and thus to thrust the psychotherapist into a role of "saviour" (Gutheil, 1989; Gutheil, 2005). Moreover, according to Gutheil (2005), people presenting dependent or histrionic traits, as well as a dependent personality disorder (DPD) or a histrionic personality disorder (HPD), also tend to test the boundaries of the therapeutic relationship. Indeed, the notion of "needs" in the case of DPD and of "drama and seduction" for those suffering from HPD can manifest in these people an intense need for contact, esteem, and approval in order to relieve their anxiety (Gutheil, 2005). The psychotherapist can succumb, progressively, to the pressure exercised by his client and transgress, or even violate the boundaries of the therapeutic relationship (Glickhauf-Hugues and Chance, 1998; Gutheil, 2005).

In a more nuanced manner, Norris et al. (2003) indicate that therapeutic touch should be reconsidered in cases where (1) there is a change in roles within the therapeutic relationship: that is, when the client tends to give complete power to the psychotherapist or, to the contrary, tries to dominate the relationship; (2) the client consults due to a recent trauma during which his boundaries were transgressed or even violated; (3) the client tends to be severely auto-critical and easily feels shame; and (4) the client is in the therapeutic relationship with the goal of living intense emotions, and tends to confuse the intimacy of the relationship with love.

Finally, Warnecke (2011) specifies that we must pay particular attention, as psychotherapists, to the risk of collusion with the client, which often leads to a loss of objectivity towards them, and that we must weigh the pros and cons when a client asks to be touched. Otherwise, we risk avoiding confronting clients on this and, in turn, disappointing them, when disappointment is sometimes also part of a therapeutic intervention (Asheri, 2009).

In What Context Does Therapeutic Touch Occur?

According to Zur (2007), many elements need to be considered when we discuss the therapeutic context. There is the context itself, the therapeutic factors, as well as the factors linked to the therapeutic relationship. Specifically, Zur (2007) indicates that the type of practice (e.g., private practice, hospital, public agency), location of practice (e.g., rural, urban, university community, home practice), the presence of other services within the same proximity, and the presence or absence of a dual relationship represent the variables linked to the therapeutic context, which need to be considered during psychotherapeutic interventions, including therapeutic touch.

In similar fashion, the methods or type of therapy, individual, couple, family, short-, medium- or long-term, the frequency and intensity of meetings, the clientele type as well as the theoretical orientation guiding interventions are among those listed by Zur (2007) as therapeutic factors.

Finally, the quality and the nature of the therapeutic alliance (e.g., secure, intense, engaged, and neutral), the duration as well as the period in which the psychotherapy is taking place (e.g., early or late in the day), the nature of the therapeutic relationship (e.g., transferential, familiar, equal, respectful), the presence or absence of a duality of roles, as well as the nature of said duality, represent the therapeutic relationship factors (Zur, 2007). Together these variables influence and mold the therapeutic framework by "the respective subjectivity of two entities that meet" (Gutheil and Gabbard, 1998, p. 412). Notwithstanding the above, the theoretical orientation in which the psychotherapy falls remains the therapeutic context variable most likely to influence the manner in which psychotherapists broach the notion of boundaries and, as such, position themselves as to the use of therapeutic touch with their clients (Borys and Pope, 1989; Gutheil and Gabbard, 1993, 1998; Lazarus and Zur, 2002; Williams, 1997).

Within the clinical supervision context, fewer guidelines are offered in regards to the use of touch between supervisor and supervisee (for more, see the Bilodeau chapter on supervision). Angus and Kogan (2007) specify that the goals, tasks, and the nature of the link between the supervisor and supervisee are all very distinct elements from the therapeutic context. As such, the supervisor's objective must first be centered on the learning and professional development of the student (Bernard and Goodyear, 2014). In addition, supervisors have the responsibility to know themselves, and to be able to evaluate their intentions and values, and to assess how these influence their use of touch in a supervision context. They must also be able to discuss openly this use with the supervisee so that, of common accord, they may decide how to include it within supervision (Kertay and Riviere, 1993; Westland, 2011). Finally, the supervisor must be aware of the imbalance of power that exists within the supervisor-supervisee relationship, the confusion that can be generated from the various roles each have and to be attentive to them, to be aware of the particulars of the person being supervised and their personal characteristics, which could interfere with the use of touch in a supervision context.

For a Better Practice of Therapeutic Touch

Guidelines and Recommendations

According to Bonitz (2008), there are few agreed-upon guidelines that regulate the use of touch within a psychotherapy framework. However, certain ethical principles should be considered in the decision process related to the use of therapeutic touch with a client. Among these principles found in the Canadian Psychological Association Code of Ethics, we find (1) treating the subject of human dignity, especially the statements concerning the free and informed consent of clients to psychotherapeutic services; (2) the principle of responsible care, specifically the statements that speak to psychotherapist competency and self-knowledge, as well as maximizing benefits while minimizing harm to the client; (3) the principle of integrity in relationships, namely the statements related to what is right; and (4) the absence of conflict of interests (e.g., absence of coercion) within the therapeutic relationship. Pope and Keith-Spiegel (2008) offer more questions and tools, allowing the psychotherapist to be more aware of professional ethics before choosing to use therapeutic touch or not. Among others, they invite the psychotherapist to (1) conceive professional ethics as being a continuous and active process that implies a constant questioning and full engagement; (2) know the ethic codes associated with the acquired accreditations, as well as the law to inform and guide the decision process; (3) be familiar with the theory and research that exists in line with the use of touch within a psychotherapy context, to develop critical thinking and reflection in regards to conclusions from such research, without accepting them from the outset; (4) observe and question oneself (e.g., Is there another way to intervene in this scenario? What happens if I am wrong? Can I meet the client's need another way?); (5) remain open to questioning of some of one's behaviours and decisions by colleagues in order to make better ethical and clinical decisions; (6) put into question the practices that one masters, or finds comfortable, rather than question the merit of practices one does not; (7) accept that ethical issues are inherent to clinical work; and (8) do not hesitate to consult colleagues who, in large part, are helpful.

Pope and Keith-Spiegel (2008) go even further and present guidelines to follow, or at the very least to consider, when the psychotherapist is called to manage and work with existing boundaries within a therapeutic relationship. They recommend that the psychotherapist (1) imagines the best and worst scenario if they transgress a boundary; (2) be familiar with scientific documentation around interpersonal boundaries; (3) be familiar with the resources offered by associations and professional colleges offering guidelines in managing boundary transgressions. For example, the College of Psychologists of Ontario offers a document treating the management of professional boundaries within the mental health field (CPO, 1998), as well as a document on the prevention of sexual abuse of clients and the obligation to report abuse (CPO, 1999). The same is true for the College of Registered Psychotherapists of Ontario; (4) identify a colleague worthy of trust to obtain feedback in regards to questions related to the transgression of boundaries; (5) be aware of one's feelings of confusion and doubt in order to better understand what may have instigated them and how these influence decision-making; (6) broach the nature of the psychotherapy in which the client is engaging from the onset of meetings (informed consent); (7) refer the client, if needed, if one is not in a proper position to treat the client with the use of therapeutic touch; and finally, (8) note all transgressions of boundaries while taking care to describe them in a factual manner, using one's clinical judgment while writing progress notes.

Likewise, the context as well as the notions of intentionality and significance become central to all psychotherapists who use therapeutic touch (Durana 1998; Holub and Lee, 1990; Hunter and Struve, 1998; Kertay and Rivere, 1993; Zur, 2007). When the psychotherapist resorts to an intervention implicating touch, they must rest this decision on a clear justification servicing the client's best interests rather than those of the psychotherapist (Westland, 2011). The psychotherapist must be able to explore their motivations to use touch with their clients and use it in a minimal and therapeutic way, that is utilizing low-risk interventions (e.g., a handshake) (Westland, 2011). Also, it goes without saying that the client must be able to consent in an informed and non-coercive way to the intervention and be encouraged by the psychotherapist, if pertinent, to share the meaning touch reveals for them (Westland, 2011). Lastly, Westland (2011) specifies that the psychotherapist must be comfortable with therapeutic touch and be trained and supervised, if necessary, in its use.

In an ideal world, all these considerations would be understood, absorbed, and used with the greatest of diligence by a majority of psychotherapists who use therapeutic touch. However, the different sections of the current chapter allow us to see that there is still a long road ahead to get there. In fact, there are many grey areas which deserve more attention within clinical practice, professional ethics, and scientific research in the therapeutic touch field.

Training, Supervision and Consultation

Caldwell (2002) indicates that the use of touch is a neglected aspect within psychotherapist training, even though it is greatly used. As such, students could greatly benefit from training that, in addition to focusing on ethics, the paradigms, and the therapeutic and cultural considerations of touch, would also emphasize their fears, their limits, and their notions of transference and counter-transference related to this intervention. Norris et al. (2003) add that concept and practice of interpersonal boundaries must also be the core of the training of future psychotherapists. More precisely, Gutheil and Gabbard (1998) mention that boundaries should be taught from the perspective of the protection of clients and the respect of the therapeutic framework, and not be presented as a list of prescribed behaviours. Along the same lines, in a supervision context, the supervisee should have the opportunity to review the therapist-client dynamic as well as the way in which interpersonal boundaries are managed, without mentioning the importance of the opportunity to discuss their counter-transference with the supervisor, a qualified professional (Norris et al., 2003). Finally, Norris et al. (2003) highlight the importance of consulting with colleagues, for psychotherapists who are already making use of therapeutic touch, when questions and/or difficulties emerge and boundaries risk being or are being compromised.

From a professional ethics point of view, according to Gutheil and Gabbard (1998), the rules designated for the protection of clients should systematically broach and consider the "therapeutic context" to determine if a transgression or violation of client boundaries occurred. While ultimately the evolution of professional ethics in therapeutic touch influences regulation by only considering norms and laws that ban this behaviour, and without consideration of all other related variables, we consent to put aside the human aspect as well as the nuances of an intervention that, "combined with a solid training, an engagement to continuous training, a humility, a sharpened clinical judgment, a refined intuition can prove to be immensely effective" (Imes, 1998, p. 198).

In summary, therapeutic touch is not straightforward. It implies many considerations, sometimes technical, but mostly human. Beyond the psychotherapist and the client lie two separate beings for whom intention and significance of touch do not have the same reach. Obviously, not knowing when faced with a professional misconduct implicating these types of paradigms is inadmissible. For this reason, it is more than necessary to better train and provide tools for accredited psychotherapists and future psychotherapists as to the use (or not) of therapeutic touch and all that it involves.

In terms of impacts and direction for future research, while the beginning of the current chapter broached the notion of physical touch, as defined by Juhan (1987), other research could have a different interest in touch. For example, the concept of self-disclosure by the psychotherapist within a therapeutic context, while already the subject of some chapters (Gutheil, 2005; Zur, 2007), could be studied in light of what Harrison et al. (2012) insist on, namely the perspective that clients have on therapeutic touch. Furthermore, research informs clinical practice and ethics which, in turn, notably influences not only the regulation but also our understanding of certain phenomenon of interpersonal boundaries. These dynamic influences then stimulate the research which, for example, could look to study psychotherapists who systemically violate their client's boundaries. Also, as previously mentioned (Imes, 1998), simply refusing to use therapeutic touch with clients may be problematic. Finding healthy and appropriate approaches to touch in helping relationships is a goal.

References

American Association for Marriage and Family Therapy. (2015). Code of ethics. Retrieved from http://www.aamft.org/iMIS15/AAMFT/Content/legal_ethics/code_of_ethics.aspx

American Counselling Association. (2014). Code of ethics. Retrieved from http://www.counseling.org/knowledge-center/ethics

Ainsworth, M. D. S. (1989). Attachments beyond infancy. *American Psychologist, 44*, 709–716.

American Association for Marriage and Family Therapy. (2012). Code of Ethics. Retrieved from http://www.aamft.org/imis15/Content/Legal_Ethics/Code_of_Ethics.aspx

American Psychiatric Association. (2013). *Diagnostic and statistical manual of mental disorders* (5th Ed.) – *DSM-5*. American Psychiatric Association: Washington.

Angus, L. and Kagan, F. (2007). Empathic relational bonds and personal agency in psychotherapy: Implications for psychotherapy supervision, practice, and research. *Psychotherapy: Theory, Research, Practice, Training, 44*, 371–377.

Asheri, S. (2009). To touch or not to touch: A relation body psychotherapy perspective. In L. Hartley (Ed.), *Contemporary body psychotherapy, the Chiron approach* (pp. 106–120). London: Routledge.

Association Canadienne de Counselling et de Psychothérapie (2007). Code de déontologie. Retrieved from http://www.ccpa-ccp.ca/_documents/CodeofEthics_fr_new.pdf

Bernard, J. M., and Goodyear, R. K. (2014). *Fundamentals of clinical supervision* (5th ed.). Upper Saddle River: Pearson.

Bowlby, J. (1969). *Attachment, separation and loss* (Vol. 1). New York: Basic Books.

Bonitz, V. (2008). Use of physical touch in the "talking cure": A journey to the outskirts of psychotherapy. *Psychotherapy Theory, Research, Practice, Training, 45*(2), 391–404.

Borderman, A., Freed, D. W. and Kinnucan, M. T. (1972). Touch like me: Testing an encounter group assumption. *Journal of Applied Behavioral Science, 8*, 527–533.

Borys, D. S. and Pope, K. S. (1989). Dual relationships between therapist and client: A national study of psychologists, psychiatrists, and social workers. *Professional Psychology: Research and Practice, 20*, 283–293.

Bosanquet, C. (1970). Getting in touch. *Journal of Analytical Psychology, 15*(1), 42–57.

Boyesen, G. (1976). The primary personality and its relationship to the streamings. In D. Boadella (Ed.), *In the wake of Reich* (pp. 81–98). London: Coventure.

Breuer, J. and Freud, S. (1957). *Studies in hysteria (1893-1895)*. New York: Basic Books.

Caldwell, C. (2002, June). *Using touch in psychotherapy*. Paper presented at the USA Body Psychotherapy Conference, Baltimore.

Carroll, R. (2002). Biodynamic massage in psychotherapy: Re-integrating, re-owning and re-associating through the body. In T. Staunton (Ed.), *Body psychotherapy* (pp. 78–100). London: Routledge.

Cohen, S. S. (1987). *The magic of touch.* New York: Harper and Row.

Cornell, W. F. (1997). Touch and boundaries in transaction analysis: Ethical and transferential considerations. *Transactional Analysis Journal, 37*(1), 30–37.

Cornell, W. F. (1998). *Touch and boundaries in transactional analysis: Ethical and transferential considerations.* In Conference Proceedings of the USA Association for Body Psychotherapy, Creating Our Community, Boulder, Colorado.

Davis, W. (2001). Energetics and therapeutic touch. In M. Heller (Ed.), *The flesh of the soul: The body we work with. Selected papers of the 7th Congress of the European Association of Body Psychotherapy 2-6 September 1999, Travamünde* (pp. 59–80). Bern: Peter Lang.

Durana, C. (1998). The use of touch in psychotherapy: Ethical and clinical guidelines. *Psychotherapy, 35*, 2369–280.

Eiden, B. (1998). The body in psychotherapy, the use of touch in psychotherapy. *Self and Society*, 26(2), 3–41.

Gabbard, G. O. (1994). Teetering on the precipice: A commentary on Lazarus's "How certain boundaries and ethics diminish therapeutic effectiveness". *Ethics & Behavior*, 4, 283–286.

Gabbard, G. O. (1996). Lessons to be learned from the study of sexual boundary violations. *American Journal of Psychotherapy, 50*, 311–322.

Galton, G. (2006). *Touch Papers, Dialogues on Touch in the Psychoanalytic Space.* London: Kamao.

Glickhauf-Hugues, C. and Chance, S. (1998). An individualised and interactive object-relations perspective on the use of touch in psychotherapy. In E. W. L. Smith, P. R. Clance and S. Imes (Eds.), *Touch in psychotherapy: Theory, research, and practice* (pp. 153–169). New York, NY: The Guilford Press.

Goodman, M. and Teichner, A. (1988). To touch or not to touch. *Psychotherapy, 25*(4), 492–500.

Gutheil, T. G. (2005). Boundary issues and personality disorders. *Journal of Psychiatric Practice, 11*(2), 88–96.

Gutheil, T. G. and Brodsky, A. (2011). *Preventing boundary violations in clinical practice.* Guilford Press: New York.

Gutheil, T. G. and Gabbard, G. O. (1993). The concept of boundaries in clinical practice: Theoretical and risk-management dimensions. *American Journal of Psychiatry,* 150, 188–196.

Gutheil, T. G. and Gabbard, G. O. (1998). Misuses and misunderstanding of boundary theory in clinical and regulatory settings. *American Journal of Psychiatry, 155*, 409–414.

Gutheil, T. G. (1989). Borderline personality disorder, boundary violations, and patient-therapist sex: Medicolegal pitfalls. *American Journal of Psychiatry, 146*, 597–602.

Harrison, C., Jones, R. S. P. and Huws, J. C. (2012). "We're people who don't touch": Exploring clinical psychologists' perspectives on their use of touch in therapy. *Counselling Psychology Quarterly, 25*(3), 277–287.

Heller, M. C. (2007). The golden age of body psychotherapy in Oslo: From vegetotherapy to nonverbal communication. *Journal of Body, Movement and Dance in Psychotherapy, 2*(2), 81–94.

Heller, S. (1997). *The vital touch.* New York: Henry Holt and Company.

Horton, J. A., Clance, P. R., Sterk-Elifson, C. and Emshoff, J. (1995). Touch in psychotherapy: A survey of patients' experiences. *Psychotherapy: Theory, Research, Practice, Training, 32,* 443–457.

Hubble, M. A., Noble, F. C. and Robinson, E. E. (1981). The effects of counselor touch in an initial counseling session. *Journal of Counseling Psychology, 28,* 533–535.

Hunter, M. and Struve, J. (1998). *The ethical use of touch in psychotherapy.* Thousand Oaks, CA: Sage.

Imes, S. (1998). Long-term clients' experience of touch in Gestalt therapy. In E. W. L. Smith, P. R. Clance and S. Imes (Eds.), *Touch in psychotherapy: Theory, research, and practice* (pp. 170–200). New York: The Guilford Press.

Jacoby, M. (1986). Getting in touch and touching in analysis. In N. Schwartz-Salant and M. Stein (Eds.), *The body in analysis* (pp. 109–126). Wilmette, IL: Chiron Publications.

Jaffy, P. (2002). Somatic Tracking and the Ethical Use of Touch. *USA Psychotherapy Body Journal, 1*(2), 63–70.

Johnson, S. (2011). Emotionally focused therapy in action. Retrieved from www.psychotherapy.net/video/johnson-emotionally-focused-therapy

Johnson, S. (2008). *Hold me tight: Seven conversations for a lifetime of love.* New York: Little, Brown and Company.

Johnston, S. H. and Farber, B. A. (1996). The maintenance of boundaries in psychotherapeutic practice. *Psychotherapy, 33,* 391–402.

Juhan, D. (1987). *Job's body.* New York, NY: Station Hill Press.

Kertay, L. and Reviere, S. L. (1993). The use of touch in psychotherapy: Theoretical and ethical considerations. *Psychotherapy, 30,* 32–40.

Knable, J. (1981). Handholding: One means of transcending barriers of communication. *Heart and Lung, 10,* 1106–1110.

Koocher, G. P. and Keith-Spiegel, P. (1998). *Ethics in psychology: Professional standards and cases* (2nd ed.). New York: Oxford University Press.

Lakin, M. (1991). *Coping with ethical dilemmas in psychotherapy.* New York: Pergamon Press.

Langs, R. (1982). *Psychotherapy: A basic text.* New York: Aronson.

Lawry, S. (1988). Touch and clients who have been sexually abused. In M. Hunter and J. Struve (Eds.), *The ethical use of touch in psychotherapy* (pp. 201–210). New York: Guilford Press.

Levine, P. A. (2008). *Healing trauma: A pioneering program for restoring the wisdom of your body.* Sounds True of Boulder, Colorado.

Levitan, A. and Johnson, J. (1986). The role of touch in healing and hypnotherapy. *American Journal of Clinical Hypnosis, 28,* 218–223.

Liss, J. (1974). Why touch? *Energy and Character, 5*(2), 1–8.

Lowen, A. (1975). *Bioenergetics.* Harmondsworth: Penguin.

Lynch, M. A. (1978). The prognosis of child abuse. *Journal of Child Psychology and Psychiatry and Allied Disciplines, 19,* 175–180.

McNeely, D. A. (1987). *Touching, body therapy and depth psychology.* Toronto: Inner City Books.

Milakovich, J. (1998). Differences between therapists who touch and those who do not. In E. Smith, P. R. Clance and S. Imes (Eds.), *Touch in psychotherapy, theory, research and practice* (pp. 74–92). London: Guilford Press.

Mintz, E. E. (1969). Touch and the psychoanalytic tradition. *Psychoanalytic Review,* 56C, 365–376.

Montagu, A. (1986). *Touching, the human significance of the skin.* New York: Perennial, Harper and Row.

Norris, D. M., Gutheil, T. G. and Strasburger, L. H. (2003). This couldn't happen to me: Boundary problems and sexual misconduct in the psychotherapy relationship. *Psychiatric Services, 54*(4), 517–522.

Older, J. (1982). *Touch is healing.* New York: Stein and Day.

Orbach, S. and Carroll, R. (2006). Contemporary approaches to the body in psychotherapy. Two psychotherapies in dialogue. In J. Carrigall, H. Payne and H. Wilkinson (Eds.), *About a body: Working with the embodied mind in psychotherapy* (pp. 63–82). London: Routledge.

Ordre des Psychologues du Québec (2014). Code de déontologie. Retrieved from https://www.ordrepsy.qc.ca/fr/psychologue/obligations/code-le-deontologie.sn

Pattison, J. E. (1973). Effects of touch on self-exploration and the therapeutic relationship. *Journal of Counselling and Clinical Psychology, 40*(2), 170–175.

Phelan, J. E. (2009). Exploring the use of touch in the psychotherapeutic setting: a phenomenological review. *Psychotherapy Theory, Research, Practice, Training, 46*(1), 97–111.

Pope, K. S., Tabachnick, B. G. and Keith-Spiegel, P. (1987). Ethics of practice: The beliefs and behaviors of psychologists and therapists. *American Psychologist, 42,* 993–1006.

Pope, K. S. and Keith-Spiegel, P. (2008). A practical approach to boundaries in psychotherapy: Making decisions, bypassing blunders, and mending fences. *Journal of Clinical Psychology: In Session, 64*(5), 368–652.

Reich, W. (1961). *The function of the orgasm.* New York: Farrer, Straus, Giroux (Original work published in 1942).

Rutter, P. (1989). *Sex in the forbidden zone: When a man in power-therapists, doctors, clergy, teachers, and others-betray women's trust.* New York: Fawcett Crest.

Schultz, L. G. (1975). A survey of social worker's attitudes and use of body and sex psychotherapies. *Clinical Social Work Journal, 3,* 90–99.

Shaw, R. (2003). *The embodied psychotherapist.* New York: Brunner-Routledge.

Showell, J. (2002). Touch in the right hands. *Psychotherapy in Australia, 8*(2), 46–50.

Simon, R. I. (1989). Sexual exploitation of patients: How it begins before it happens. *Psychiatric Annals, 19,* 104–122.

Smith, D. and Fitzpatrick, M. (1995). Patient-therapist boundary issues: An integrative review of theory and research. *Professional Psychology: Research and Practice, 26,* 499–506.

Smith, E. (1985). *The body in psychotherapy.* Jefferson, NC: McFarland.

Smith, E. (1998). Traditions of touch in psychotherapy. In E. Smith, P. R. Clance and S. Imes (Eds.), *Touch in psychotherapy, theory, research and practice* (pp. 3–15). London: Guilford Press.

Société Canadienne de Psychologie (2000). Code canadien de déontologie professionnelle (3ième Ed.). SCP/CPA.

Sonne, J. L. (1994). Multiple relationships: Does the new ethics code answer the right questions? *Professional Psychology: Research and Practice, 25,* 331–343.

Southwell, C. (1988). The Gerda Boyesen method: Biodynamic therapy. In J. Rowan and W. Dryden (Eds.), *Innovative therapy in Britain* (pp. 178–201). Milton Keynes: Open University Press.

Stake, J. E. and Oliver, J. (1991). Sexual contact and touching between therapist and client: A survey of psychologists' attitudes and behavior. *Professional Psychology: Research and Practice, 22,* 297–307.

Staunton, T. (2000). Sexuality and body psychotherapy. In T. Staunton (Ed.), *Body psychotherapy* (pp. 56–77). Hove: Brunner-Routledge.

Stenzel, C. L. and Rupert, P. A. (2007). Psychologists' use of touch in individual psychotherapy. *Psychotherapy: Theory, Research, Practice, Training, 41,* 332–345.

St. Germaine, J. (1996). Dual relationships and certified alcohol and drug counselors: A national study of ethical beliefs and behaviors. *Alcoholism Treatment Quarterly, 14*(2), 29–45.

Strasburger, L. H., Jorgenson, L. and Sutherland, P. (1992). The prevention of psychotherapist sexual misconduct: Avoiding the slippery slope. *American Journal of Psychotherapy, 46,* 544–555.

Strozier, A. L., Krizek, C. and Sale, K. (2003). Touch: Its use in psychotherapy. *Journal of Social Work Practice, 17,* 49–62.

The College of Psychologists of Ontario. (2009). Normes de conduite professionnelle. Retrieved from http://www.cpo.on.ca/Ressources_et_publications.aspx.

The College of Psychologists of Ontario. (1999, August). Prevention of sexual abuse of clients and mandatory reporting. *The Bulletin, 26*(1), 1–2.

The College of Psychologists of Ontario. (1998, July). Professional boundaries in health-care relationships. *The Bulletin, 25*(1), 1–5.

Tirnauer, L., Smith, E. and Foster, P. (1996). The American Academy of Psychotherapists Research Committee survey of members. *Voices, 32*, 87–94.

Toronto, E. L. K. (2006). A clinician's response to physical touch in the psychoanalytic setting. *International Journal of Psychotherapy, 7*(1), 69–81.

Totton, N. (2003). *Body psychotherapy*. Philadelphia: Open University Press.

Totton, N. (2005). *New dimensions in body psychotherapy*. Maidenhead: Open University Press.

Tune, D. (2005). Dilemmas concerning the ethical use of touch in psychotherapy. In N. Totton (Ed.), *New dimensions in body psychotherapy* (pp. 70–84). Maidehead: Open University Press.

Warnicke, T. (2011). Stirring the depths: transference, countertransference and touch. *Body, Movement and Dance in Psychotherapy, 6*(3), 233–243.

Wen-Shing, T. and Streltzer, J. (Eds.). (2001). *Culture and psychotherapy: A guide to clinical practice*. Washington, DC: American Psychiatric Press, Inc.

Westland, G. (2011). Physical touch in psychotherapy: Why are we not touching more? *Body, Movement and Dance in Psychotherapy, 6*(1), 17–29.

Williams, M. H. (1997). Boundary violations: Do some contented standards of care fail to encompass commonplace procedures of humanistic, behavioral, and eclectic psychotherapies? *Psychotherapy, 34*, 238–249.

Wilson, J. M. (1982). The value of touch in psychotherapy. *American Journal of Orthopsychiatry, 52*, 65–72.

Young, C. (2007). The power of touch in psychotherapy. *International Journal of Psychotherapy, 11*(3), 15–24.

Zur, O. (2007). *Boundaries in psychotherapy: Ethical and clinical explorations*. Washington, DC: American Psychological Association – APA Books.

Zur, O. and Nordmarken, N. (2009). To touch or not to touch: Exploring the myth of prohibition on touch in psychotherapy and counseling, clinical and ethical. Retrieved from www.zurinstitute.com/touchintherapy.html

Conclusion

Issues of Touch: An Overall View and Integration

Martin Rovers, Judith Malette and Manal Guirguis-Younger

Touch is an extremely significant human experience and perhaps one of the most susceptible to interpretation. A handshake could be one of friendship, or of peacemaking after a conflict, of greeting and welcoming someone new, to seal a promise or an agreement, to congratulate a victory, and finally to say goodbye. The context dictates which of these meanings attaches itself to the same handshake. It would seem a handshake is not a simple gesture after all, and derives its significance or lack thereof from the circumstances in which it is carried out. Touch lies at the heart of our human development, of our experience of self, of our encounter with the world; and naturally touch comes into play in the helping relationships, like counselling and spiritual care. Yet it remains difficult to discuss touch as part of the human dynamic in the helping relationship and is left out of clinical conversation. There are many reasons that can explain this: the therapist's theoretical orientation and values around touch which can dictate his/her level of comfort with touch in the therapeutic context; numerous complex issues regarding touching a vulnerable individual; professional boundaries and ethical concerns around touching; and risk management, to name a few. Unfortunately, the lack of conversation has also led to the lack of clear guidelines from professional colleges around issues of touch. In this chapter, we wish to summarize some of the themes covered in this book on touch, and to highlight some areas of touch in counselling and spiritual care. We are thankful to all those who wrote chapters in this volume, and gave us so many perspectives on the role and place of touch in helping relationships. Contributions from academic fields such as anthropology, philosophy, spirituality and psychology, from individual to couple therapy, nursing, as well as training and supervision, have given us a comprehensive conceptualization of the impact of touch on service providers and recipients alike. In this chapter, we wish to summarize these main contributions, and to highlight some areas of touch in counselling and spiritual care that were not specifically treated in this volume, yet remain important to consider.

Touch Helps Us Thrive

Most importantly, this book aims to highlight that the study of human development, and especially attachment theory, has empirically supported touch as essential for our emotional and relationship growth (Bowlby, 1988). Over the past 50 years, great research strides were made in the area of the psychology of touch and its relationship to attachment. Harlow's research on our need for "contact comfort," Bowlby's work showing that touch is a necessary aspect of our human nature, Montague's breakthrough research on touch, and much more (see Chapter I on touch deprivation), all direct us to a necessary full-court review on the place of touch in the helping professions, and the depth of encounter between client and therapist, nurse and patient, and student and supervisor. The philosophy, psychology, and biology of touch have changed our way of thinking over the years. We are tactile beings. Secure attachment is built upon early and healthy touch. Montague (1971) states that children's hands and bodies begin to relax within seconds of touch. Touch is vital to achieving a sense of emotional and psychological health and decreasing one's feeling of isolation. When we touch, we thrive.

Field (1987) researched infants in hospital settings to examine how touch impacts their development. In one of her most important studies (1987), she found that massage therapy leads to 45% weight gain in premature infants versus control groups of non-massaged infants. The infants who were massaged also leave the hospital six days earlier than non-massaged infants. When following up with the infants one year later, the massaged infants were still at a weight advantage and were now scoring higher on tests of mental state and motor skills. Here, research led to massage therapy becoming an integral part of premature infant care in many hospitals.

Similar results of improved health have been found in other studies as well, for both the elderly and sick. Eaton, Mitchell-Bonair and Freidmann (1986) found that elderly residents of long-term health care facilities, who were given an encouraging touch more frequently, ate larger amounts at meal time. When touched frequently, they consumed both more calories and proteins, which are key components in overall physical health. Improved health was also found to be linked to touch in an HIV study (Diego, Hernandez-Rei, Field, Friedman and Shaw, 2001). The study followed teens diagnosed with HIV for one month. During this month, the teens received massage therapy. Compared with the control group, the massaged teens were found to have an increase in natural killer cells that help fight the symptoms of the disease. Individuals of all ages have been found in countless studies to reap many psychobiological benefits. Touch improves physical and emotional health. All of these studies strongly suggest that touch is essential to life.

Touching You; Touching Me; Touching Puppies; Touching in the Helping Relationships

Touch in counselling and spiritual care has been much debated and written about, with no clear consensus about how to actually talk about it or bring up the topic within the helping relationship setting itself. Yes, we all know the necessity and value of human touch for social and especially emotional growth and healing, but practitioners seem to be clearly told, "Do not touch!" Apparently, many people long to be touched in a loving manner, but everyone is afraid to do any touching in the helping relationships and indeed afraid even just to talk about it. It seems as if even bringing up the topic of touch with clients/patients is wrong or tempting suggestion.

Before practitioners can use touch within the helping relationship, they need to talk about touch with their clients/patients. What is touch? When and how it might be brought into this professional relationship and for what purpose? And how can both practitioner and client/patient mutually consent to the possibility of touch within their relationship (Westland, 2011)? "The omission of tactile experience is noticeable not only in the field of history, but across the humanities and social sciences. It seems that we have so often been warned not to touch that we are reluctant to probe the tactile world even with our minds" (Classen, 2012, p. x1).

The challenge of touch is that when you touch someone, they are touching you. When done well, touch is mutually beneficial. Touch is healing, and yet, in the helping relationships, we are so fearful to think about or initiate touch. Research shows that the majority of counsellors believe that touch and physical contact is beneficial for patients and clients (Joshi, Almeida and Shete, 2010). "The most commonly considered benefits of touch were that it helps to calm the client, strengthen the therapeutic bond, communicate acceptance and respect, facilitate self-disclosure and catharsis of emotions" (p. 41). Various forms of touch are frequently used in the helping relationship. Pope, Tabachnick and Keith-Spiegel (1987) found that 41% of psychologists within the American Psychological Association reported they hugged their clients. Milakovich (1998) interviewed 84 therapists and found that 70% of them used touch in their private practice. Stenzel and Rupert (2004) found that 80% of therapists report some kind of touch in their relationships with clients, even if only the touch of greeting and termination. Harrison, Jones, and Jaci (2012) found that "the potential for touch to lead to change and its description is a 'pivotal moment'" (p. 286). Touch is used more in therapy with children and in spiritual care and chaplaincy, and is a common and necessary part of nursing, medicine, and physiotherapy. Touch is vital to achieving a sense of emotional and psychological health and decreasing one's feeling of isolation.

Yet it seems that few counsellors and helping relationship practitioners incorporate touch into their practice, or even have a conversation with their clients about the use or place of non-erotic touch. Female practitioners and older counsellors tended to have a more positive attitude towards touch and used it more in their relationships with clients (Chillot, 2013). We all know the necessity and value of human touch for social and especially emotional growth and healing.

Most writers on touch in psychotherapy determine that, if touch is to be used in psychotherapy, clients/patients need to be informed of the touch, the type of touch, and the rationale for using touch (Smith, 1998; Hunter and Struve, 1998; Corey, Corey and Callanan, 1998; Westland, 2011). Williams, Clarke and Gibson (2011), in one of the more up-to-date research studies on touch in psychotherapy, found that there is still divergence in terms of touch in therapy with some, usually older clinicians and more often female clinicians, tending to touch more, while others never touch at all. They also found that touch was scarcely talked about in academic courses and training in psychotherapy since the days of Hunter and Struve (1998) or Orbach (2006). Indeed, other than greeting and leaving moments of therapy, little touch was done. In short, this research found that fear was the dominant theme: fear that touch would be misinterpreted as sexual, fear that the therapist might be breaching their own professional ethics, and fear of litigation.

There are many kinds of touch: welcome greeting and handshake touch; accidental touch when you brush against another; task orientated touch, in terms of helping someone; celebratory touch, the high-five; consoling touch and grieving; prayerful touch such as the laying on of hands; affectionate touch in friendship; emotional and sexual touch; and abusive or aggressive touch (Hunter and Struve, 1998). Some greeting and grieving touches are done almost naturally in many helping relationship encounters. Children and most clients have little trouble touching puppies, or in canine-assisted therapy, bringing a sense of ease and trust again. Others need explicit permission. Berendsen's chapter on Somatic Experiencing therapy uses experiential touch: the therapist asks the client to do something, like gesture anger, as a means towards healing. A new, consented touch is used to reorientate clients away from old wounding touch to new safe touch. Touch is used in Satir's family therapy sculpturing. Interpartner touch is incorporated in couple counselling suggested by PACT and EFT (see Petrella and Rovers in Chapter VII). Touch is what the helping relationship is about, namely doing what is best for the client in a sensitive, permissible manner.

Gelb (1982) identified four factors associated with clients' positive evaluation of touch in psychotherapy: (1) clarity regarding the boundaries of therapy; (2) congruence of touch;)3) client perception of being in control of the physical contact; and (4) the client's perception that touch is for his/her benefit rather than the practitioner. Horton, Clance, Sterk-Elifon and Emshoff (1995) added two more: (1) the degree of therapeutic alliance; and (2) the potential for sexual attraction in the therapeutic dyad. All these factors need to be considered if and when touch is introduced in the helping relationship. Above all, explicit client consent is required.

When we consider that some forms of touch usually happen in therapy (handshake, accidental touch, consoling touch), and research supports the helpfulness of positive touch within the therapeutic relationship, it seems a logical follow-up that professionals need to talk and contract touch in their relationship. This should be done right from the beginning when other ethical matters are clarified and agreed upon, like the limits of confidentiality and the benefits and potential harm of counselling. Open contractual conversations on touch allow a rich opportunity to explore expectations, feelings, self-perceptions, and interpersonal issues that can be evoked by a conversation of touch, thus enabling clients and therapists to reach a mutually agreeable understanding from the beginning. Too often, however, there is no conversation about touch, and awkward moments happen when two people, especially one with power over the other, touch and both know they are on different pages of life's acceptable manual, and there are misinterpretations, or worse, violated boundaries. Too often, there is or has been no contractual conversation or agreement.

Contracting Touch in the Helping Relationships

One of the intentions of this edited book is to open the dialogue about touch in the therapeutic relationship and seek ways in which we can talk about touch, even at the contractual level, in the beginning of the therapeutic relationship. In the client contract, we talk about duty to warn and to protect, limitations of confidentiality, the benefits and hindrances to therapy, and fees. Touch deserves and demands its place at these contractual conversations. So how do we contract touch in counselling and spiritual care? Consent is a prerequisite whenever touch may be introduced into the therapeutic relationship. There are different types of consent, best defined along the line of different times and places where consent can be obtained. Consent can be (1) obtained at the time just before touch is introduced into the therapy relationship; (2) secured up front as part of the contractual agreement (see Hunter and Struve, 1998); or (3) assumed: consent that is acceptable and appropriate in the general public, like greeting touch.

Whichever approach is used for touch within a helping relationship, effort needs to be taken, before engaging in any expression of touch, to verbalize and clarify specific boundaries and guidelines that any use of touch might have in the situation. Principles of openness, transparency, and mutual affirmation need to be considered, as well as clear boundaries that touch is within the power of the client/patient. For example, even at the first moment of therapy, like a handshake of welcome, this can be clarified by a simple, "Are you okay with a handshake?" Hunter and Struve (1998) suggest that although they do not usually use such an informed consent form, they offered one in their book. This informed consent form is used as a part of their normal intake process, although such written, upfront consent is used little in practice and suggested little by others. Caution is needed; such discussion on possible touch at some point in therapy can cause insecure and vulnerable clients/patients more fear than good, and thus open up more questions than answers. Talking about touch can also be confusing and create possible misinterpretation, and at times seem to be too suggestive.

Contracting consent for touch can be done on a case-by-case basis, making the decision at the time when touch might be appropriate. Thus, touch might be introduced once the relationship has been established and the client/patient can feel more comfortable with the therapist and the therapeutic process. In other words, get permission to touch as needed for solid therapeutic reasons (Durana, 1998). Some examples are: "I notice that you are shaking a lot; can I touch your arm as a sign of reassurance?" and "You asked that I pray with you, and I am happy to do this. Can I touch your hand as we pray?"

Touch within the psychotherapeutic process needs to be done with huge consideration of transparency, power of decision with the client/patient and clarity of boundaries, and a clear contract and/or understanding, openly talked through. Above all, there needs to be open discussion about touch, allowing the client/patient and psychotherapist space to explore what touch might mean to them. Constantly checking about the meaning of any contact is also important. Psychotherapists need to decide about the use of touch with each client/patient on an individual basis.

There are advantages as well as difficulties in an upfront conversation about touch in the helping relationship. Having the conversation means both parties learn more about each other, their style of communication, comfort in the world with others, boundaries, possible attachment styles, and comfort with self and others. It is clear and written.

Clients Who Want to Touch or Be Touched

The case can arise where the client is dealing with a loss or a difficult moment and asks the therapist for a hug, or the client greeting or hug becomes too prolonged. Of course, touch can be desired by the client but be inappropriate for the therapeutic moment, or uncomfortable for the therapist. Perhaps the client's/patient's culture is accustomed to touch more than the therapist's own experience/cultural expectations of touch. Therapist intuition and experience, theoretical orientation, or understanding of the client's dynamics can be the best judge of appropriate touch when asked by the client/patient. What happens when the therapist wants to or needs to say no to touch? How does the therapist do this in a clear yet therapeutically affirming manner?

Different cultures and individuals have varying degrees of tolerance for touch, and touch can mean vastly different things to different people; it is thus wise to clarify all of this as the therapeutic relationships begin. It is much harder later in the therapeutic relationship to correct possible misinterpretation of the touch language, such as a client/patient asking you for a hug, but your therapeutic and ethical principle is never to hug your client patients, and you then have to say no, and the client/patient may think, "What a cold therapist"; or the client/patient just expressed a lot of grief and seeks some consolation, which is most often done by touch. Touch or lack thereof always sends emotional signals: It is one of the more primary ways of communication between people. This is no different in the therapeutic relationship. Indeed, Chillot (2013) suggests that we can communicate and decode emotions via touch alone, and that we are "wired to interpret the touch of our fellow humans" (p. 34).

The Ethics of Touch

In our book, we specifically arranged a chapter on the ethics of touch in the helping relationships for the simple reason that, beyond research that clearly tells us about the vital necessity of touch in the lives of us all, we want to clearly articulate that within the various helping relationship professions, such as counselling, spiritual care, nursing, and supervision, touch must always be done within clear and well-defined boundaries. For touch to be healing, abundant knowledge about the power of touch, for good and for bad, needs to be balanced with great sensitivity for those who are our clients.

Yet there is still the strong sense within the therapeutic and spiritual care communities that touch is inappropriate and, worse, unethical, almost to the point of not talking about it. Any coverage about touch in training courses strongly suggests that it needs to be avoided, or at least minimal, like a handshake or touch on the arm. What might touch mean within the helping relationships, but equally, what does the possible absence of touch say to our clients? What message might professionals be giving clients when touch is not done, or not even talked about, in terms of the client's self-worth, or healing processes?

To touch or not to touch in psychotherapy needs a more expansive conversation than "No!" And therefore the question of touch in psychotherapy has mostly been a short chapter in the ethics section of texts that can curtly be summed up in a similar fashion: "Don't do it." It is not the intention of this book on touch to really change the cautions regarding touch or the ethics thereof in psychotherapy. We are fully aware of and adhere to the professional and ethical prescription and guidelines of how inappropriate touch in the psychotherapist-client relationship can be an abuse of power and devastating for both client and professional. Anderson's (1999) somewhat less stringent guiding principle for touch is still sound: touch should be minimal, if at all, and stick to the safe zone of handshakes, high-fives, and shoulders or arms. The ethics chapter of this book expands the ethical connotations of touch in the helping relationships.

What Touch Might Mean Within a Therapeutic Relationship

In their chapter, White and McPherson write that touch is "a highly complex, intersubjective and context-dependent act that has meaning for those involved" (p. 245). As mentioned by Demasure in Chapter IV, as well as at the beginning of the epilogue, touch is the only sensory modality that is reciprocal, that is it is impossible to touch without being touched. This reciprocity is evocative of being present to the other and to oneself, and vice versa. It also speaks of an encounter taking place between an "I" and a "you." From an existential perspective, this encounter involves maximal closeness while keeping a respectful distance from the phenomenological field of the other person. It involves empathy, compassion, trust, and faith in oneself and in the other. "I" becomes in contact with the "You" (Buber, 2002/1947). "'I-Thou' touch involves one's whole being touching another whole being" (Westland, 2011, p.18). Whether touch involves a physical contact through skin, eye contact, or even a contact through smell, it means two individuals are relating and touching each other emotionally, intellectually, and even spiritually. A patient sharing the narrative of a painful episode in her life once said, "For many years after the abuse took place, I could not tolerate being touched, not even with a gentle touch. I have come to realize that although I was trying to protect myself, I was also punishing myself. Not being touched or not touching means not being in relation with other people. I want to trust, even though this is not easy for me. I want to live and feel worthy of touching and being touched." Analyzing this narrative goes beyond the scope of the epilogue. However, touch, even a gentle touch, is never a neutral act. It carries with it profound existential meaning, which does not mean touching in a professional way should be a forbidden act either. As psychotherapists, we may not know the exact meaning touch does carry for our clients, but it is our responsibility to be mindful and respectful of this way of relating to and encountering the other.

Touch as Presented Within Academic Education and Training

Most training programs mention very little about touch within the helping relationships, and when it is mentioned, it is usually to tell students not to do it at all, as a bad thing, even though research suggests it can be helpful at the right time and in the right way (Hunter and Struve, 1998). Older (1982) found that this negative attitude towards touch in psychotherapy results in touch being used significantly less in psychotherapy than in the other helping relationships, like spiritual care, chaplaincy, nursing, occupational therapy, and medicine. Have we become too cautious at the cost of losing vital aspects of the human connection?

Conclusion

Were we to ask psychotherapists out of the blue if they touch their clients, the vast majority would say, "No!" Yet we can see that, as we delve a bit further into the reality of touch within the helping relationships, our first instinctual answer of "no" has room for further clarification and discussion, perhaps moving practitioners to begin to think about the ways "yes" to touch are possible. Most helping relationship practitioners seem to be comfortable with various gestures of touch that are socially appropriate. Most helping relationship professionals also seem to subscribe most to contracting touch on an as-needed basis, bringing up the topic when it is appropriate. Written contracts and conversations about touch right up front in session seem to be a much more delicate situation, and require great caution, as mentioned before. That might just be the nature of touch within any relationship that is based on secure attachment and trust. It would be a foolish potential mate who approaches the first date with the question, "How can I touch you?" When we place the question of touch within a context of a secure and safe attachment within the helping relationship, and touch as an important element of the encounter with another, then the rules of attachment theory need to be our guide. There needs to be, first, a secure and trusting relationship. There needs to be plenty of talk, trust, and reliability before touch, as appropriate, can be a more used aspect of therapy to help people and relationships heal.

References

Anderson, P. A. (1999). *Nonverbal communication: Forms and functions*. Mayfield Publications.

Buber, M. (2002). *Between man and man*. London: Routledge Classics. (Original work published in 1947).

Chillot, R. (2013). The power of touch. *Psychology Today*. March, 2013.

Classen, C. (2012). *The deepest sense: A cultural history of touch*. Chicago: University of Illinois Press.

Corey, G., Corey, M. S. and Callanan, P. (1998). Issues and ethics in the helping relationship. Pacific Grove, CA: Brookes/Cole Publishing.

Downey, D. L. (2001). Therapeutic touch in psychotherapy. *Psychotherapy, 36*(1), 35–38.

Durana, C. (1998). The use of touch in psychotherapy: Ethical and clinical guidelines. *Psychotherapy Theory, Research, Practice, Training, 35*, 269 280.

Goldschmidt, B and van Meines, N. (2012). *Comforting touch in dementia and the end of life care*. London: Singing Dragon.

Harrison, C., Jones, R. S. P. and Huws, J. C. (2012). We're people who don't touch: Exploring clinical psychologists' perspectives on their use of touch in therapy. *Counselling Psychology Quarterly, 25*(3), 277–287.

Horton, J. A., Clance, P. R., Sterk-Elifson, C. and Emshoff, J. (1995). Touch in psychotherapy: A survey of patients' experiences. *Psychotherapy, 32*(3), 443–457.

Hunter, M. and Struve, J. (1998). *The ethical use of touch in psychotherapy*. Thousand Oaks: Sage Publications.

Joshi, P. D., Almeida, M. and Shete, P. D. (2010). Attitudes toward physical contact in the therapeutic setting: Role of gender and expertise. *Journal of the Indian Academy of Applied Psychology, 36*(1), 35–43.

McGuirk, J. (2012). *The place of touch in counselling and psychotherapy and the potential for healing within the therapeutic relationship*. Retrieved from: http:/iahip.org.

Milakovich, J. (1998). Differences between therapists who touch and those who do not. In E. W. L. Smith, P. R. Clance and S. Imes (Eds), *Touch in psychotherapy: Theory, research, and practice* (pp. 74-91). New York: The Guilford Press.

Montagu, A. (1986). *Touching: The human significance of the skin*. New York: Harper.

Older, J. (1982). *Touch is healing*. New York: Stein & Day.

Osbach, S. (2006). Too hot to touch? In G. Galton (Ed.), *Touch papers; Dialogues on touch in the psychoanalytic space* (pp. xiii–xviii). London: Karnac.

Smith, E. W. L., Clance, P. R. and Imes, S. (1998). *Touch in psychotherapy: Theory, research, and practice*. New York: The Guilford Press.

Thayer, S. (1982). Social touching. In W. Schiff and E. Foulkes (Eds.), *Tactual perceptions*. Cambridge: Cambridge University Press.

Westland, G. (2011). Physical touch in psychotherapy: Why are we not touching more? *Body, Movement and Dance in Psychotherapy*, 6(1), 17–29.

White, K. (2004). *Touch, attachment, and the body*. Karnac Books.

Williams, S., Clarke, D. and Gibson, K. (2011). *The use of touch in counselling and psychotherapy*. LAP Lambert Academic Publishing.

Contributors

Christian Bellehumeur, PhD, C.Psych, is an associate professor of the Faculty of Human Sciences and Philosophy at Saint Paul University and the director of the School for Counselling, Psychotherapy and Spirituality at Saint Paul University. He is a member of l'Ordre des psychologues du Québec. His research focuses on the imaginary and human development as well as the links between psychology, religion and spirituality.

Patricia Berendsen, MA, is a psychotherapist in private practice in London, Ontario, and specialized in somatically based psychotherapy. As a Somatic Experiencing Practioner, Patricia incorporates touch as an integral component of her clinical practice. Her interest in trauma and the body has resulted in numerous peer-reviewed publications. Additionally, Patricia is an accomplished speaker and workshop facilitator. She is a consultant to agencies and provides training for organizations that wish to incorporate a trauma-informed and somatic perspective into their work.

Cynthia Bilodeau, PhD, Ed, is a registered psychotherapist and an assistant professor in the Faculty of Human Sciences and Philosophy at Saint Paul University. Her areas of research relate to clinical supervision and professional training. She is also interested in the role of individual factors such as shame, anxiety, well-being and personality in the processes of professional development. Cynthia has published several national and international research articles related to her areas of interest and has authored two personality inventories.

Jane Chambers, MA, is a doctoral candidate in counselling and spirituality at Saint Paul University. She is dedicated to accompanying adults living with the consequences of childhood trauma, as well as people struggling with their gender and sexual identity. She is presently exploring chronic illness narratives through the lens of Durand's theory of anthropological structures of the imaginary, in order to understand the experiences of people who live with chronic illness.

Nathalie Charron, MA, is a counsellor and therapist in private practice in addition to offering immediate emotional support within an employee assistance program. Her interests include sexuality, feminism, boundaries in the therapeutic relationship, transference/counter-transference, trauma, and personality disorders.

Isaac Davis, MA, is a registered psychotherapist with the College of Registered Psychotherapists of Ontario. He completed his undergraduate degree at Carleton University in the Bachelor of Arts, Honours Program, in Psychology, and holds a Master's degree in counselling and spirituality from Saint Paul University. Isaac operates a private practice in Kanata, Ontario, working with individuals, couples and families. Some of his major focuses include helping teens establish an identity, helping parents to communicate effectively with their teenager, and partner and intimacy related issues.

Karlijn Demasure, PhD, is a professor in practical theology teaching at the Gregorian University in Rome (Italy). She is the director of the Center for Child Protection. Previously, she was the president of the Interdiocesan Service for Pastoral Care in Belgium. She held the Chair in Christian Family Studies at Saint Paul University in Ottawa, Canada, where she also was the dean of the Faculties of Human Sciences and Philosophy. She has been an invited professor at several universities. Her field of research is sexual abuse, trauma, spirituality, and care.

Richard Feist, PhD, is an associate professor of the Faculty of Human Sciences and Philosophy at Saint Paul University. His research interests include ethics, metaphysics, phenomenology, philosophy of science and mathematics, and process philosophy (especially Alfred North Whitehead). A frequent guest speaker at conferences around the world, he also teaches and writes on human knowledge, moral philosophy, process metaphysics and God, and military and peacekeeping ethics.

Marilyn Guindon, PhD, C. Psych, is an associate professor at the School of Counselling, Psychotherapy and Spirituality of the Faculty of Human Sciences and Philosophy at Saint Paul University. She is also a clinical psychologist. Her main clinical and research interests focus on schema therapy for trauma, spiritual experience and anxiety disorders.

Manal Guirguis-Younger, PhD, C.Psych, is a professor of counselling and the dean of the Faculties of Human Sciences and Philosophy at Saint Paul University. Dr. Guirguis-Younger's research is focused in the areas of community homelessness, palliative care, volunteer resources and alternative models of health service delivery to marginalized groups. She is interested in the unique health and social needs of marginalized individuals, especially at the end of life. Her work provides frontline solutions aimed to overcome the challenges of service provision to groups with complex health issues. Dr. Guirguis-Younger is the editor-in-chief of the book *Homelessness and Health in Canada* (University of Ottawa Press, 2014). She continues her work on the role of volunteers in palliative care and has recently been a co-editor of *Le bénévolat en soins palliatifs ou l'art d'accompagner* (Presses de l'Université Laval, 2013).

Nazila Isgandarova, DMin, has a degree in pastoral counselling, marriage and family studies from Wilfrid Laurier University and a Master of Social Work degree from the University of Windsor. She is a registered social worker (RSW) with the Ontario College of Social Workers and Social Service Workers (OCSWSSW) and a Registered Psychotherapist with the College of registered psychotherapists of Ontario (CRPO). She works as Spiritual and Religious Care coordinator at the Ontario Multifaith Council and Spiritual Care Provider at the Centre for Addiction and Mental Health, and provides psychotherapy in a private practice. She also has authored several articles that have appeared in different academic books and journals.

Kristine Lund, PhD, lives in Waterloo, Ontario, and is a professor at Waterloo Lutheran Seminary (WLS) and holds the Albert J. Datar Chair in Spiritual Care and Psychotherapy. She has also published in the area of professional ethics, teaching, and learning. Kristine is the happy owner of a Portuguese Water Dog named Annie, who participates in animal-assisted activities at WLS and at the Delton Glebe Counselling Centre.

Judith Malette, PhD, C.Psych, is a full professor in the Faculty of Human Sciences at Saint Paul University in Ottawa. She is also a clinical psychologist with 16 years of experience. Her areas of research are images of God and images of self, life review, end of life issues, yoga and psychospiritual growth. Judith is the author, with Marilyn Guindon, of "Les images de Dieu chez les personnes ayant été abusées sexuellement," in K. Demasure (Ed.) (2014). *S'élever au-delà de l'abus sexuel.* (pp. 31–51), Brussels: Lumen Vitae, in the "Care and Spirituality" series; and, with Manal Guirguis-Younger, of "L'utilisation d'une technique narrative, la Relecture de Vie, en milieu palliatif." In A. Sévigny, M. Champagne, and M. Guirguis-Younger (Eds.) (2013). *Le bénévolat en soins palliatifs ou l'art d'accompagner* (pp. 239–253), Maison Michel Sarrazin and Presses de l'Université Laval. In the past two years, she has written and taught on compassion fatigue at the Pontifical Gregorian University in Rome (e.g., Burnout and Its Variations: The Necessity for Self-Care, Pontifical Gregorian University [Rome]), Center for Child Protection: Safeguarding: Our Commitment (CCP E-Learning Programme), Unit 18: Self-Care for Helping Professions.

Linda Mayorga Miller, MA, is a chaplain at The Ottawa Hospital. She is a Canadian Association for Spiritual Care certified specialist in spiritual care. Her research interests include the use of self-generated ritual and personal sacralized objects as coping and hoping strategies. She has published an article and presented numerous workshops in this area.

Christine McPherson, PhD, is a health psychologist and registered nurse who acquired her doctorate from the Institute of Palliative Care and Policy at King's College, London (UK). She is an associate professor in the School of Nursing at the University of Ottawa, where she teaches research and palliative care at the graduate level. Her research spans over 15 years and includes more than 80 publications and presentations aimed at understanding and improving the care of patients at the end of life and their family members.

Reesa Packard, MA, is a doctoral candidate in counselling and spirituality at Saint Paul University. She is a clinical therapist and researcher who works with individuals and couples to assist them in their journeys towards healing and self-actualization. In both research and practice her specialties include childhood and relational traumas, yoga and spiritual development and relational connectedness.

Cassandra Petrella, MA, is a psychotherapist in clinical practice. She completed her degree with a specialization in couples counselling. She is the owner and lead psychotherapist of Senta Counselling Services in Ottawa. She specializes in working with couples, women's issues, anxiety and chronic pain management. She is currently researching inter-partner touch with Professor Martin Rovers at Saint Paul University.

Martin Rovers, PhD, C.Psych, is a professor, CASC CPE Supervisor, and AAMFT Approved Supervisor in the Faculty of Human Sciences at Saint Paul University in Ottawa. He has been a health care practitioner for over 30 years, working as an educator, psychologist, marriage and family therapist, team leader, mentor, writer, and workshop presenter. Martin is on the executive of The Society for Pastoral Counselling Research. Martin is the author of *Healing the Wounds in Couple Relationships* (Novalis, 2005), *The Helping Relationships: Healing and Change in Community Context* (Co-editor, University of Ottawa Press, 2010), and *Family's Many Faces: Contemporary Family Patterns and Challenges* (2014: Co-editor: Peeters, Belgium), as well as many articles in the area of attachment within couples and family. He has developed Attachment in Family Therapy (AFT), which is a synthesis of attachment theory and family or origin theory. AFT uses a PLANTING Hope model for marriage and family counselling. Currently, Martin is coordinator of the Couple and Family program with the Masters of Counselling, Psychotherapy, and Spirituality at Saint Paul University. He is married to Elizabeth, and they have a daughter, Paulina.

Lacie White, RN, PhD (c), CHPCN (c), is a doctoral candidate in nursing at the University of Ottawa, and a visiting student researcher and doctoral fellow at the University of Victoria. As a palliative care nurse,

Lacie is dedicated to expanding the depth and breadth of practice knowledge and education in the field. Her research focuses on ways in which palliative care nurses can embody being present and mindful amidst the complexities of experiences and relationships inherent at end of life.

Health and Society

Series editor: Sanni Yaya

The Health and Society series provides a space for dialogue where different fields of expertise (sociology, psychology, political science, biology, nutrition, medicine, nursing, human kinetics, and rehabilitation sciences) generate new insights into health matters from the individual as well as the global perspectives on population health.

Previous titles in this collection

Manal Guirguis-Younger, Ryan McNeil and Stephen W. Hwang (editors), *Homelessness & Health in Canada*, 2014

www.press.uottawa.ca